A BEGINNER'S GUIDE TO PLANETS A

Other Worlds

TERENCE DICKINSON

ILLUSTRATIONS BY
DAVID EGGE AND JOHN BIANCHI

FIREFLY BOOKS
BOOKMAKERS PRESS

Canadian Cataloguing in Publication Data

Dickinson, Terence
Other worlds

Includes index.
ISBN 1-895565-71-5 (bound)
ISBN 1-895565-70-7 (pbk.)

1. Planets — Juvenile literature. I. Title.

QB602.D53 1995 j523.4 C95-931046-0

Front Cover: Early stages of the formation of a planetary system around a newborn star. Illustration by David Egge.

Back Cover: A brown dwarf about 10 times the mass of Jupiter. Illustration by David Egge.

Published by
Firefly Books
250 Sparks Avenue
Willowdale, Ontario
Canada M2H 2S4

Published in the U.S. by
Firefly Books (U.S.) Inc.
P.O. Box 1338
Ellicott Station
Buffalo, New York 14205

Produced by
Bookmakers Press
12 Pine Street
Kingston, Ontario
K7K 1W1

Design by
Linda J. Menyes
Q Kumquat Designs

Color separations by
Chroma Graphics (Overseas) Pte. Ltd.
Singapore

Printed and bound in Canada by
Metropole Litho Inc.
St. Bruno de Montarville, Quebec

Printed on acid-free paper

To Blair and Adrienne, who will find new worlds in the 21st century

NOTE TO READERS
Some of the artists' renderings in this book might be mistaken for spacecraft photographs —and vice versa. If in doubt, consult the photo/illustration credits listing on page 63.

CONTENTS

4
WORLDS IN SPACE

6
THE PLANETS AND THEIR MOONS

8
OVERVIEW OF THE UNIVERSE

10
ORIGIN OF THE PLANETS

12
THE EARTH'S MOON

14
MARS

16
MARTIAN LANDSCAPES

18
VENUS

20
MERCURY

22
ASTEROIDS

24
JUPITER

26
IO

28
EUROPA

30
GANYMEDE AND CALLISTO

32
SATURN

34
TITAN

36
OTHER MOONS OF SATURN

38
URANUS

40
NEPTUNE

42
TRITON

44
PLUTO

46
THE KUIPER BELT

48
COMETS

50
COMET CRASH ON JUPITER

52
BROWN DWARFS

54
OTHER PLANETARY SYSTEMS

56
PULSAR PLANETS

58
PLANETARY EXPLORATION

59
THE SUN AND ITS PLANETS

60
MOONS OF THE PLANETS

62
SOURCES

63
THE AUTHOR AND CREDITS

64
INDEX

Earth and its Moon are two worlds among a family of 9 planets and 61 known moons in our solar system. Trillions of unknown worlds likely exist around other stars.

When I was about 9 years old, I discovered that other worlds exist. Since this was during the early 1950s, before most homes had television, including mine, the only place I could see another world—another planet—was in a book. That's how I made my discovery.

In school one day, I happened to open the classroom atlas, a huge book occasionally used by my teacher. Near the front of the atlas was a small section about the universe. My attention was instantly fixed on an illustration on one of those pages, a painting that showed the nine planets of the solar system strung out in a row. For the first time, I realized we knew the *sizes* of these other worlds that orbit the Sun along with Earth. They had dignified names such as Neptune, Mercury and Jupiter. But I remember something else too: The surfaces of these worlds—all of them except Earth—were almost blank. Just smudges or vague cloud belts appeared.

How things have changed! Now, not only do we know exactly what the planets look like close up, but we know what these worlds and their moons are made of and the temperature and composition of their atmospheres. We have measured their sizes to an accuracy of a few kilometers. We have mapped and named mountains on Venus, canyons on Mars and giant valleys on Saturn's moon Tethys. Today, we can predict with confidence what it would be like to stand on any of the 9 planets and their 61 known moons—or at least on those that have solid surfaces!

Virtually everything we have learned about the planets and their moons has been gathered during the quarter-century beginning about 1965, a period astronomer Carl Sagan has called "the golden age of planetary exploration." In that span—slightly longer than one human generation—space probes have visited every planet except Pluto. Specially equipped landers have sent back images from the surface of Mars and of Venus. And almost every month now, the Hubble Space Telescope turns its penetrating gaze toward our neighbor worlds, the planets and their moons.

Sometime during the 21st century, humans will set foot on the orangish sands of Mars. In the more distant future, explorers will reach Pluto, the ice world at the rim of the solar system. There, when they turn away from the Sun, these interplanetary astronauts will look toward the stars beyond and see them as the next frontier.

But exploring the planets and traveling to the stars are two very different enterprises. If the Sun were the size of a baseball sitting on home plate in Yankee Stadium in New York City, Earth would be about the size of the ball in a ballpoint pen lying in the grass eight meters (26 ft) away. Pluto, the size of a grain of sand, would be in the last row of seats in the stadium, while the other planets would be scattered between there and home plate. But the nearest star, Alpha Centauri, would be in Winnipeg, Manitoba. And the next nearest star would be in Tucson, Arizona.

Travel to the stars is clearly out of the question for the near future, but there are billions of stars out there that could have planets orbiting them—possibly even a world like Earth. Later in this book, I will discuss the search for such planets in detail. But to prepare for the voyage ahead, the next four pages offer a compact overview of our own solar system and its place in the universe.

LEFT **Viewing other worlds from Earth. This photograph shows our celestial neighbor, the Moon, lined up with Venus, the planet closest to Earth. The more distant planet Jupiter is seen at upper right.**
RIGHT **When the Galileo spacecraft passed Earth on its way to Jupiter in December 1992, it captured this image of Earth and the Moon from a distance of six million kilometers (4 million mi). Floating in the blackness of space, the two worlds are a study in contrasts: the Moon, a completely barren, airless desert, and Earth, mostly covered by water and teeming with life.**

THE PLANETS AND THEIR MOONS

In this illustration, the Sun, its planets and all their known moons are shown at correct scale size. Of course, accurate scale distances between objects cannot be depicted. To do so would require a sheet of paper 15 kilometers (9 mi) across. The Sun and Mercury, for example, would be 60 meters (200 ft) apart. For data on any object shown here, see the tables on pages 59-61.

SUN

JUPITER'S MOONS
1. Metis
2. Adrastea
3. Amalthea
4. Thebe
5. Io
6. Europa
7. Ganymede
8. Callisto
9. Leda
10. Himalia
11. Lysithea
12. Elara
13. Ananke
14. Carme
15. Pasiphae
16. Sinope
URANUS'S MOONS
1. Cordelia
2. Ophelia
3. Bianca
4. Cressida
5. Desdemona
6. Juliet
7. Portia
8. Rosalind
9. Belinda
10. Puck
11. Miranda
12. Ariel
13. Umbriel
14. Titania
15. Oberon
URANUS
SATURN'S MOONS
1. Pan
2. Atlas
3. Prometheus
4. Pandora
5. Janus
6. Epimetheus
7. Mimas
8. Enceladus
9. Tethys
10. Telesto
11. Calypso
12. Dione
13. Helene
14. Rhea
15. Titan
16. Hyperion
17. Iapetus
18. Phoebe
SATURN
NEPTUNE'S MOONS
1. Naiad
2. Thalassa
3. Despina
4. Galatea
5. Larissa
6. Proteus
7. Triton
8. Nereid
NEPTUNE
Charon
PLUTO

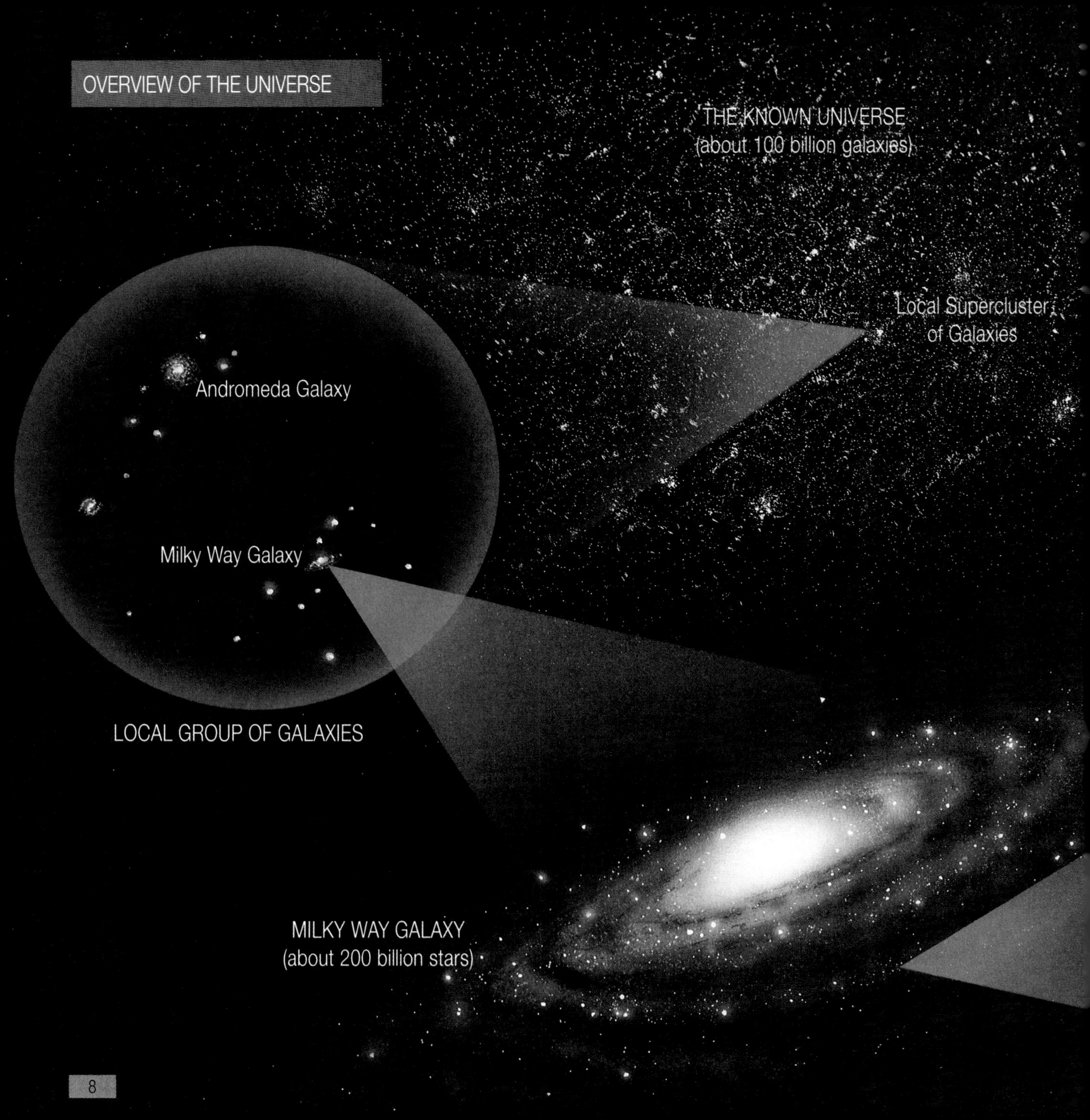

THE KNOWN UNIVERSE
(about 100 billion galaxies)

Local Supercluster
of Galaxies

Andromeda Galaxy

Milky Way Galaxy

LOCAL GROUP OF GALAXIES

MILKY WAY GALAXY
(about 200 billion stars)

INNER SOLAR SYSTEM
Sun
Mercury
Earth
Venus
Mars
Asteroid Belt
OUTER SOLAR SYSTEM
Uranus
Jupiter
Saturn
Pluto
Neptune
Alpha Centauri
Sun
NEARBY STARS

Stars form when enormous clouds of gas and dust collapse. Planets are born from a disk of matter that surrounds the new star.

Imagine a cloud of gas so vast, a beam of light would take hundreds of years to travel from one side to the other. A thimbleful of this cloud, which is far more diffuse than the thinnest of the Earth's clouds, would contain just a few hundred atoms and molecules. Compare that with a thimbleful of the air you are now breathing, which contains a million trillion atoms and molecules. But as diffuse as it is, this giant cloud has enough matter to create thousands of stars.

Star formation occurs when a section of such a cloud begins to contract, or pull itself inward by its own gravity. This process may be triggered by a nudge from a shock wave from a nearby exploding star. In a few million years, the cloud has contracted until it is several times wider than the orbit of Neptune, the outermost large planet in our present solar system.

At this point, the cloud has begun to spin and shape itself into a vast disk. At the heart of the disk, matter falls inward, creating a seething ball of hot gas. As more matter falls in toward the center, pressures and temperatures rise. Eventually, the temperature is high enough and the pressure great enough to ignite the gas in a hydrogen-bomb-type nuclear reaction. This marks the birth of a star like the Sun.

Meanwhile, cosmic dust in the cloud material in the disk around the new sun has begun clumping into celestial "dust balls." As these fluffy balls bump into each other, they stick together, forming larger balls. Soon, monster dust clumps as big as a house are circling the newborn sun. The process of clumping continues, until millions of tiny planets are whirling around the star, each in its own orbit. These mini-planets, called

CENTER **Known as the Orion Nebula, this cloud of gas and cosmic dust is an enormous star factory, where thousands of stars are being born. Around many of those newly forming stars, planets are probably being created as well.**

LEFT **When the Sun was born about 4.7 billion years ago, it was surrounded by a thick ring of gas, dust and chunks called planetesimals. After millions of years, the planetesimals collected into larger objects, which would become the planets.**

RIGHT **This illustration shows Earth and the Moon during a period of intense bombardment by asteroids and comets. The bombardment occurred from 4.5 to 3.9 billion years ago, and impact craters from this era still cover the Moon.**

planetesimals, collide with each other, sometimes smashing apart and sometimes, depending on the speed of collision, merging into a larger object.

We think the rocky inner planets of our own solar system—Mercury, Venus, Earth and Mars—were built up in just this way. During the early formation stage, the Sun's heat pushed the lighter gases into the outer solar system. Earth and its neighbors formed without atmospheres. Repeated pummeling from planetesimal impacts kept their surfaces molten for a hundred million years or more. Slowly, they cooled. Venus, Earth and Mars gained atmospheres from gases released by volcanoes, and water may have been supplied by impacts from comets, which are primarily water ice.

In the outer part of our solar system, where the Sun's heat is greatly diminished, the gas was not dispersed from the giant planets. Jupiter, Saturn, Uranus and Neptune have rocky cores roughly the size of Earth that are surrounded by thick atmospheric blankets of hydrogen, helium and other gases.

The Earth's cratered Moon is a cosmic museum. Its surface has preserved a four-billion-year record of asteroid and comet bombardment.

When astronaut Buzz Aldrin, the second human to set foot on another world, was taking his first steps on the lunar surface on July 21, 1969, he said something that I have always remembered. As he looked around at the alien landscape, he exclaimed: "Magnificent desolation!" It was a perfect description of the stark, inhospitable lunar surface.

Standing on the Moon's surface, the astronauts saw a barren vista of rocks, boulders and fine dust. In the distance, rounded hills and mountains reached toward an utterly black sky. The Moon has no atmosphere, no water, no known volcanoes and no "moonquakes" of any significance. The only substantial changes that occur on our neighbor world are caused by impacts with cosmic debris.

The thousands of craters on the Moon are evidence of huge collisions with asteroids and comets sometime in the past, but such impacts are now extremely rare. Almost all the craters were formed more than three billion years ago. Today, the objects hitting the Moon are mostly micrometeorites, tiny sand-grain-sized pieces which grind down the lunar rocks into the dust that covers the Moon's surface.

The Moon is right next door, astronomically speaking. If you take a cantaloupe and a ping-pong ball and place them 4.5 meters (15 ft) apart, you'll have a scale model of the Earth-Moon system. On this same scale, the Sun is 17 meters (56 ft) in diameter and 2 kilometers (1¼ mi) away.

Where did the Moon come from? For almost a century before the Apollo astronauts landed on the Moon, astronomers had debated the merits of three ideas:

- the adopted-cousin theory (the Moon was a small planet gravitationally captured by Earth)
- the sister theory (Earth and the Moon were born as a double planet)
- the daughter theory (the Moon split off from a spinning primordial Earth)

But detailed examination of the lunar material returned to Earth by the Apollo missions failed to support any of these ideas. A new hypothesis, which I have dubbed the chip-off-the-old-block theory, was developed with the use of modern computer simulations of the conditions that may have existed during the formation of the solar system. The simulations suggest that at the time of the birth of the Sun and planets, there may have been a few giant planetesimals, possibly as large as Mars, roaming in the region where Earth was forming.

One of these objects collided with the young Earth. The heat generated by two objects that big smashing into each other would have melted vast amounts of material from both Earth and the impacting body, and the material would have splashed into nearby space. The simulations show that a portion of this wreckage would have been captured by the Earth's gravity, creating a ring around our planet. The ring material eventually formed the Moon.

The chip-off-the-old-block theory fits with analyses of lunar material returned by the Apollo astronauts. Such studies revealed that Moon rocks contain very little iron. According to the theory, the splashed-out material that coalesced into the Moon would have come from the *surfaces* of Earth and the impacting body, not from their iron-rich cores.

Thus the Moon's origin appears to have been a fluke, rather than a natural planetary-formation process. The chip-off-the-old-block theory therefore explains why only Earth has a large moon while the other rocky planets (Mercury, Venus and Mars) do not.

LEFT **Astronaut Gene Cernan stands beside the lunar rover in this photograph from the Apollo 17 mission in 1972. His footprints show that the lunar soil is a fine, powdered rock. The Moon's surface has been ground down into this dustlike rock by billions of years of bombardment from mostly sand-sized meteorites. The mountains in the background are covered with the same material.**

RIGHT **The Moon as it appears from Earth through a telescope. The large craters are about 100 kilometers (60 mi) across.**

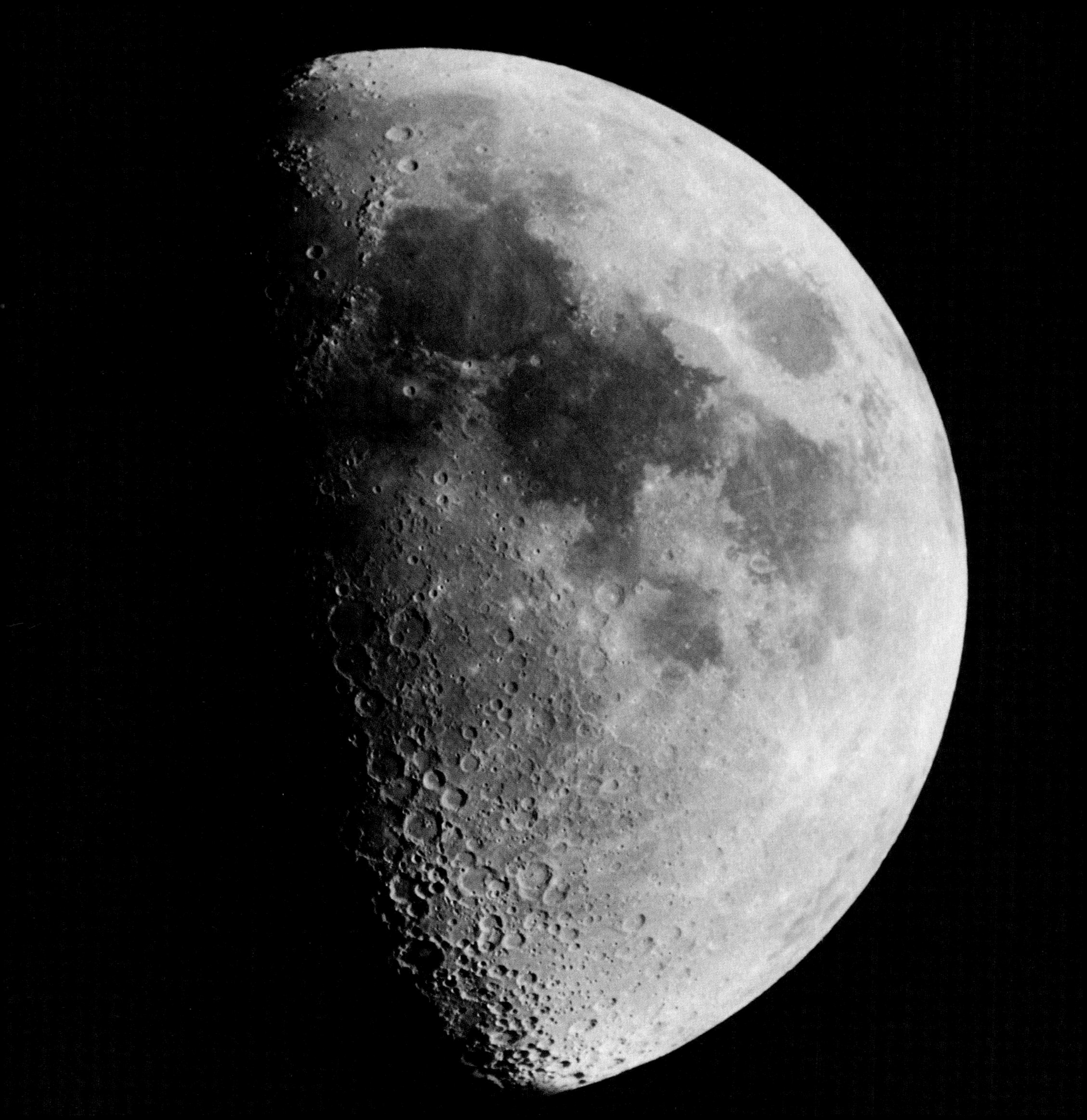

Mars is the most Earthlike of all the planets in our solar system.
But astronomers once thought it was even more like Earth than it actually is.

Mars has always been my favorite planet. My fascination with the mysterious desert world was ignited at the age of 10, when I read some *John Carter of Mars* comic books. The comics were based on a series of novels written by Edgar Rice Burroughs, who later became famous as the creator of Tarzan. The stories featured a Mars inhabited by alien creatures that Earthman John Carter always managed to outsmart by the last page.

A few years later, I graduated to the classic H.G. Wells novel *The War of the Worlds*, which described an attack on Earth by Martians. It was the Mars of Wells—a Mars of canals and Martians—that everyone born before the mid-1950s grew up with. The old Mars that Wells wrote about was fashioned more than a century ago at an observatory in Flagstaff, Arizona, where astronomer Percival Lowell peered through a 24-inch telescope at the peach-colored Martian globe and recorded what he saw as thin lines cobwebbing the planet.

From his observations, which were later shown to be incorrect, Lowell developed an elaborate theory about waterways—the canals of Mars—constructed by a dying civilization on a parched planet. So powerful were his ideas that as recently as 1965, NASA issued an official map of Mars which still had some of Lowell's canals marked on it.

Finally, the camera eyes of spacecraft from Earth revealed the true Mars—a landscape of frigid, windswept deserts that make Antarctica seem balmy by comparison. There are no thin, straight lines of the type that Lowell reported seeing on Mars, although there are canyons caused by shifting landmasses and channels that geologists say were created by flowing water.

LEFT A view across Mariner Valley, a Martian canyon 4 times deeper and 20 times wider than the Earth's Grand Canyon. The huge feature was discovered in 1972 by Mariner 9, the first spacecraft to orbit Mars. The yellow-orange sky is from wind-blown dust suspended in the atmosphere.
CENTER Potato-shaped Phobos, the larger of the two moons of Mars, is 28 kilometers (17 mi) long and 19 kilometers (12 mi) wide. Mars appears in the background in this image taken in 1989 by a Russian space probe.
RIGHT Mars, as seen by the Hubble Space Telescope in February 1995, reveals its icy north polar cap, vast deserts and wispy ice-crystal clouds. At the left edge, poking through the densest of the Martian clouds, is the giant volcano Olympus Mons, which is pictured on page 17.

Today, there are polar caps of ice but no liquid water on Mars. Billions of years ago, though, flowing water carved channels on the red planet.

Mars is midway between Earth and the Moon in size, as shown in the illustration on page 6. From that bit of information alone, we might expect Mars to have features in common with both worlds. And, as it turns out, it does. Here's a list of the chief Earthlike characteristics:

- Mars has volcanoes, deserts and dried-up riverbeds.
- Mars has an atmosphere, although it is much thinner than the Earth's.
- The rotation rate of Mars—its day—is only 39 minutes longer than an Earth day.
- The temperature at its equator reaches 0 degrees C (32°F) around midday, about the same temperature as a late-November day in Toronto or Chicago.
- Mars' rotation axis is tipped at almost exactly the same angle as the Earth's, producing the Martian equivalent of spring, summer, fall and winter.
- The polar caps on Mars are made of water ice, like the Earth's. During winter, however, the Martian polar caps are overlaid by a coating of dry ice—frozen carbon dioxide.

Balancing that list, though, is a roster of unEarthlike items:

- There is no liquid water anywhere on the surface of Mars. What water there is on the planet exists as ice at the poles or as permafrost—a cementlike mix of ice and dirt that remains solid below the surface for millions of years.
- The Martian atmosphere is *much* thinner than the Earth's. Even the air at the top of Mount Everest is 40 times denser than Mars' atmosphere.
- Large sectors of the Martian surface are covered by ancient impact craters like those on the surface of the Moon.
- The temperature at the equator on Mars drops to minus 70 degrees C (–94°F) *every night*. That's colder than Antarctica in midwinter. Away from the equator, it is colder still.

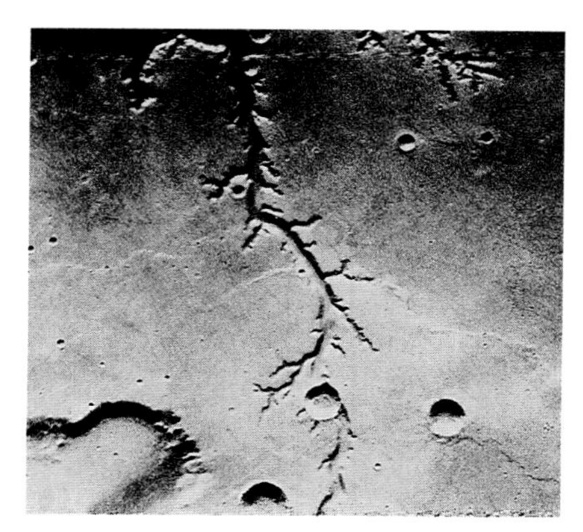

ABOVE **Water once flowed on Mars. There is no other reasonable explanation for the channel we see here, which is roughly the same width and length as the Earth's Grand Canyon. There are more than 100 similar features on the red planet.**

RIGHT **The colossal Olympus Mons (Mount Olympus) volcano on Mars is by far the tallest mountain yet observed in the solar system. It stands three times higher than Mount Everest and covers an area larger than France. It last erupted sometime between 100 and 800 million years ago. This spacecraft view shows the huge structure as it appears from directly above.**

Despite these alien attributes, Mars remains the most Earthlike world we know. Spacesuited astronauts will have no trouble walking on Mars, since its gravity is only two-fifths as strong as the Earth's. Future explorers will examine the mighty Martian volcanoes to determine when they last erupted and whether they might do so again. They will visit the polar icecaps and drill into the layers of ice to develop a better understanding of the past climate on Mars. And they will walk along the ancient channels, where water once flowed across the Martian surface.

The existence of more than 100 channels on Mars indicates that liquid water washed across at least part of the planet sometime in the past. Where there was water, there might have been life. What remains unanswered is how long the water era lasted and why it ended.

After examining the thousands of spacecraft images returned from Mars by three Martian orbiters (Mariner 9, Viking 1 and Viking 2), geologists have concluded that the water era on Mars occurred in the far distant past. Most estimates are between 3.5 and 4 billion years ago. At that time, there may have been oceans on Mars. Researchers have suggested that there was once enough water to cover the entire planet with an ocean at least 100 meters (300 ft) deep.

This more Earthlike period didn't last long. While some of the water still exists today as permafrost and polar ice, most of it either combined chemically with the soil or disappeared—Mars' low gravity could not prevent a thicker atmosphere like the Earth's from escaping into space. Today, Mars is an extremely dry, frigid desert, which would make Antarctica seem balmy.

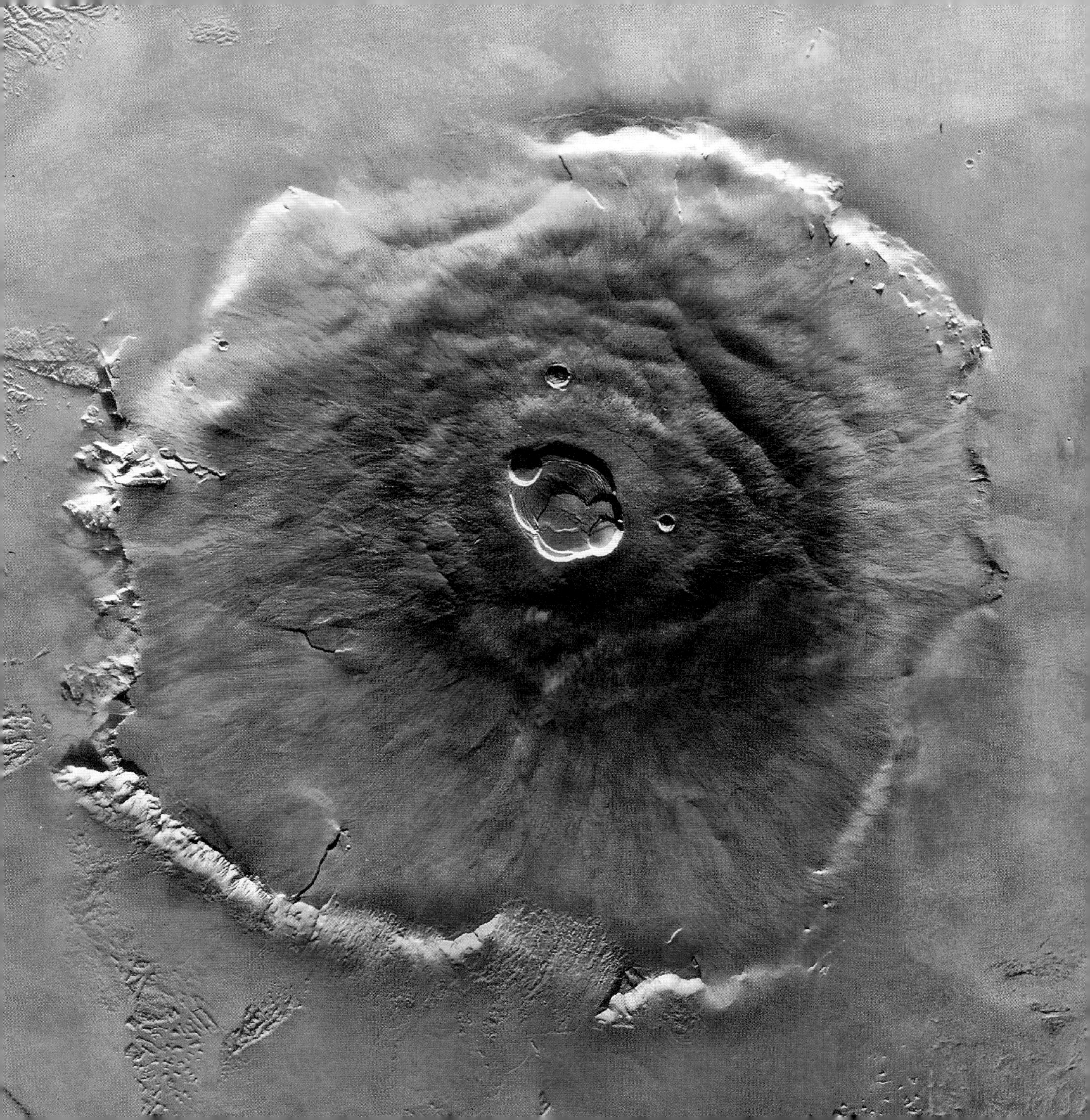

Venus may look Earthlike at a glance, but its thick, hot atmosphere makes it one of the most hostile environments in the solar system.

Of all the worlds in the solar system, only Venus comes close to matching our planet in size. In many ways, Venus could be called the Earth's twin. It is slightly smaller in diameter, and its surface gravity is 90 percent of the Earth's.

If that were the extent of the differences between Earth and Venus, then a visitor to our near neighbor world would feel right at home. But the reality is that size and surface gravity are the *only* similarities between these two. Everything else about Venus is radically different.

Standing on Venus would mean instant death for a human—or any other form of life as we know it. Venus is cloaked in an atmosphere of carbon dioxide 90 times as dense as the Earth's air. The atmospheric pressure on Venus is the same as that experienced by a deep-diving submarine on Earth 1,000 meters (3,300 ft) below the ocean's surface. In such an environment, a human body would suddenly implode.

But pressure is far from the whole story. Venus's atmosphere is made up almost entirely of carbon dioxide, which acts like a giant greenhouse, trapping heat from the Sun. This greenhouse effect has turned the surface of Venus into an inferno—hotter even than Mercury, the closest planet to the Sun. Night and day, from equator to pole, the surface rock of Venus is 460 degrees C (860°F), hot enough to melt lead.

High in Venus's atmosphere, from an altitude of about 48 to 65 kilometers (30-40 mi), lies a layer of sulfuric-acid clouds. These creamy yellow clouds, made of the same stuff we call battery acid, completely block any view of the

LEFT **The sunlit portion of Venus appears entirely featureless, because its thick atmosphere is topped by a haze of sulfuric acid and sulfur dioxide.**
CENTER **Several robot spacecraft have parachuted to the surface of Venus, which is shown in this illustration of a region of Venus that resembles the barren canyonlands that exist on Earth. Although Venus may look Earthlike, its dense atmosphere and broiling temperatures keep the planet off limits to human explorers.**
RIGHT **In the early 1990s, the American Magellan spacecraft mapped Venus's surface by radar, which was able to penetrate the cloudy atmosphere to reveal a surface of vast lava flows, volcanoes, mountains and rocky plains.**

planet's surface. Below the clouds, a sulfuric-acid mist extends down to within 31 kilometers (19 mi) of the surface. Below that, the air is as hot as the mouth of a blast furnace and is completely clear.

The Magellan space probe, which orbited Venus in the early 1990s and mapped the planet by radar, revealed that the landscape is entirely rocky plains and solidified volcanic lava flows. Enough sunlight penetrates to the surface to make the average day on Venus look like a dull, cloudy day on Earth. From the rocky surface, Venus's daytime sky is a featureless yellow haze.

Actually, Venus and Earth may have been much more alike in the distant past. Some researchers suggest that oceans emerged on Venus about the same time they did on Earth. But the added heat from the Sun at Venus's distance soon evaporated the oceans. That meant carbon dioxide released by the volcanoes on Venus could no longer be absorbed into the oceans. As a result, the carbon dioxide remained in the air to produce the deadly greenhouse effect.

Mercury is a planet of extremes. It has intensely hot days, frigid nights, icy polar regions and the shortest year and longest day of any planet.

Half a century ago, a close-up of the surface of the planet Mercury appeared on the cover of the November 1948 issue of *Scientific American* magazine. The image showed a team of astronauts standing on a hill overlooking a valley. The valley floor was broken by a network of cracks that appeared to have been caused by intense heat. Above, a brilliant white Sun glared down from a black, starry sky.

The *Scientific American* cover picture was fiction, of course—a painting by a brilliant Hollywood film artist named Chesley Bonestell. But Bonestell's Mercury painting, along with many other planetary scenes he did during the 1940s and 1950s, convinced millions of people that spaceflight to other worlds was possible. And it became a reality faster than almost anyone predicted.

In 1974, only 26 years after Bonestell painted the Mercury landscape, the American robot space probe Mariner 10 flew close to the innermost planet and radioed back to Earth hundreds of detailed photographs of Mercury's rugged

LEFT **Mercury's Caloris Basin, the large bull's-eye pattern extending off the left side of this picture, is the biggest crater-type feature on the innermost planet's surface. The smallest craters in this view are about five kilometers (3 mi) wide.**
RIGHT **A view of Mercury's cratered landscape as seen by the American spacecraft Mariner 10. At first glance, Mercury looks like the Earth's Moon, and in many ways, the two worlds are similar.**

landscape. They revealed that Mercury is covered with more craters than are found on our Moon.

Mercury, like the Moon, has an ancient surface, largely unchanged since the major asteroid and comet bombardment four billion years ago. By midafternoon in most regions of the planet, the surface rocks are toasting at 250 degrees C (480°F). Some places near the equator reach an astounding 400 degrees (750°F). At night, the surface cools to about minus 170 degrees (–275°F). But since Mercury has no atmosphere, astronauts would still be able to explore. With no air to transmit heat (as on Venus), an insulated, air-conditioned spacesuit would provide adequate protection.

Future explorers could avoid Mercury's broiling midday heat by landing near the poles, where the Sun remains on the horizon and the surface temperature is always below freezing. In 1991, scientists were surprised when radar signals sent from Earth and reflected off Mercury indicated that sheets of ice exist at the planet's poles, either on the surface or just beneath it. The polar ice could be frozen water from a comet that crashed into Mercury and exploded long ago.

Earlier radar studies in 1965 uncovered Mercury's unusual day/night rotation cycle. On this planet, the day is exactly twice as long as the year! A year on Mercury—the period it takes the planet to orbit the Sun—is 88 Earth days long. But a "day"—the interval from one sunrise to the next—is 176 days, the longest of any planet.

The gap between the orbits of Mars and Jupiter is the realm of the asteroids —a belt of millions of rocky mini-planets that are just beginning to be explored.

Anyone who has played an outer-space video game has encountered asteroids —giant cosmic boulders menacingly hurtling across the screen. Millions of these mini-planets exist. Some are no larger than a delivery van, while others are the dimensions of one of Saturn's middle-sized moons. The vast majority of asteroids orbit the Sun between Mars and Jupiter. Apparently, Jupiter's gravitational influence prevented a single large planet from forming in this region.

About 6,000 asteroids have their orbits mapped with reasonable accuracy. When plotted on a chart, the known asteroid family looks like the illustration on the facing page. Roughly 200 known asteroids stray outside the main belt zone. A few dozen of these swing inside the orbit of Earth, while a couple of others venture as far out as Saturn's orbit.

The largest asteroid, Ceres, is 915 kilometers (569 mi) in diameter, about the same width as the state of Texas. It orbits the Sun once every 4.6 years at 2.8 times the Earth's distance. The next largest asteroids are: Pallas, at a diameter of 522 kilometers (324 mi); Vesta, at 500 kilometers (311 mi); and Hygiea, at 430 kilometers (267 mi). Twenty-six asteroids are more than 200 kilometers (120 mi) in diameter. There are probably at least 100,000 over 1 kilometer (0.6 mi) in size, millions larger than a barn and billions bigger than a refrigerator.

With all that debris roaming the asteroid belt, it would seem a dangerous place. And in scores of science fiction films, that is exactly how the asteroid belt is depicted. In the *Star Wars* sequel, *The Empire Strikes Back,* for instance, the only escape route for the good guys is through a dreaded asteroid field. The hero, Han Solo, plunges his spaceship into the asteroids, jockeying it like a car in a demolition derby. But this image of asteroid belts as outer-space pinball machines, with craggy crater-pitted boulders jostling up against one another, is a total departure from reality.

In fact, there is plenty of room between Mars and Jupiter for trillions more asteroids than are already there. If we were to embark on a trip from Earth to Jupiter, passing right through the asteroid belt, we would probably see fewer than a dozen asteroids. And even those would still be so distant that they would appear to us as moderately bright stars, rather than looming boulders.

LEFT **On its way to Jupiter, the American spacecraft Galileo snapped this picture of the asteroid Ida. Unexpectedly, the image also revealed a tiny moon orbiting Ida, the first known moon of an asteroid. Ida is 52 kilometers (32 mi) long, and its moon, named Dactyl, is 1.5 kilometers (1 mi) in diameter. Both are composed of rock.**

RIGHT **More than 5,000 asteroids are plotted in this illustration of the asteroid belt. The orbits of the vast majority of the asteroids are located between 2.2 and 3.2 times the Earth's distance from the Sun. A few wander beyond the main belt: one loops out as far as Saturn, while another swings inside Mercury's orbit.**

SUN
ORBIT OF MERCURY
ORBIT OF VENUS
ORBIT OF EARTH
ORBIT OF MARS
ASTEROID BELT
ORBIT OF JUPITER

King of the planets, Jupiter is a colossal world so big that it could contain all the other planets and still have room to spare.

Imagine Jupiter as a watermelon and Earth as a grape. This comparison will give you a good idea of how huge Jupiter really is. It is the most unEarthlike of all the worlds in the solar system.

The first time I looked at Jupiter through a telescope, I was amazed. There, floating in the blackness of the eyepiece field of view, was a brilliant cream-colored disk crossed by several light brown stripes. Four tiny "stars"—Jupiter's largest moons—were strung out beside the giant planet. It's like a miniature solar system, I thought.

The idea that Jupiter has more in common with a star than a planet was emphasized by astronomer Carl Sagan when he called Jupiter a "failed star." What Sagan was referring to is that Jupiter is a gas planet, made up mostly of hydrogen and helium, and the Sun—a star—is also primarily hydrogen and helium. The difference, however, is that Jupiter is not massive enough to produce the immense pressures in its core required to "burn" its hydrogen the way the Sun does. To do so, Jupiter would have to be at least 80 times more massive than it is. (For more on this point, see page 52.)

The visible face of Jupiter consists of cloud tops—a veil of haze and ice crystals floating at the surface of an atmospheric ocean. Like the skin of an apple, the clouds seen in the picture on the facing page (and on page 31) are simply the colorful wrapping on a giant gaseous globe. Far below, encased at the very center of the giant planet, is a small rock-and-metal core the size of Earth.

Human exploration near Jupiter seems highly improbable in the foreseeable future. The planet's intense magnetic field acts like a cage around the giant planet, trapping lethal radiation belts in an invisible shield. A voyage through this radiation barrier would be deadly. Someday, however, it may be possible for humans to probe the planetary colossus using a protective submarine-type device. But for now, only robots will be doing the exploring.

Among the more surprising of the many Voyager discoveries at Jupiter was that of a very faint ring surrounding the planet. Made up mostly of dust-sized particles, the ring is exceedingly difficult to detect. It would be completely invisible to explorers standing on the surface of Jupiter's four large moons.

The dustlike ring particles probably come from the innermost small moons of Jupiter, whose surfaces are gradually worn down through constant contact with hurtling subatomic particles in the intense Jovian magnetic field. Another source could be material ejected into space by the spectacular volcanoes on Jupiter's moon Io (page 26).

ABOVE **An explorer descending into Jupiter's atmosphere near the Great Red Spot would see a colorful vista of towering windswept clouds.**

RIGHT **Jupiter's moon Io treks in front of the giant planet, casting an inky shadow on the cloud-banded panorama below. This Hubble Space Telescope image, the best ever obtained from Earth, was taken in 1994, shortly before Comet Shoemaker-Levy 9 collided with Jupiter.**

IO

Jupiter's moon Io is the most volcanically active world in the solar system. Huge eruptions constantly belch molten sulfur onto the plains of this bizarre satellite.

Late in the afternoon of March 9, 1979, the computer brain of the Voyager 1 space probe signaled its electronic cameras to swing toward Jupiter's four largest moons: Io, Europa, Ganymede and Callisto. This was the climax of Voyager's encounter with Jupiter. At Mission Control at the Jet Propulsion Laboratory in Pasadena, California, hundreds of scientists had been watching the images being returned to Earth as the space probe approached Jupiter and its family of moons. The pictures of Jupiter were showing tremendous detail. The great swirling clouds that blanket the giant planet filled the field of view in Voyager's cameras.

Now, the mission plan called for distant views of Europa, medium-range looks at Ganymede and Callisto and an extremely close pass to within 21,000 kilometers (13,000 mi) of Io (pronounced EYE-oh).

Astronomers expected all four moons to be cratered worlds similar to our Moon. But the first close-up view of Io did not reveal a single crater. Instead, this world appeared splotched and splattered with weird patches of yellow, orange and brown, unlike anything the researchers had ever seen before. One scientist said the image looked like a giant pizza rather than a celestial object. The features on Io were so bizarre and so unfamiliar that it was several days before astronomers realized what they were seeing.

Once they examined the best close-range pictures, the astonishing truth became clear: Io is pocked with giant active volcanoes! The Voyager pictures revealed several of the volcanoes spewing yellow and brown jets and umbrella-shaped plumes of molten sulfur. The jets and plumes soared hundreds of kilometers above Io's surface. When Voyager 2 passed Jupiter four months later, six of the eight volcanoes seen by Voyager 1 were still erupting. And the Voyager 2 pictures disclosed three more.

The source of this fury is Io's hot liquid interior, which acts like a pot of boiling water. Io's solid surface crust is the lid on the pot. Something has to give, and the hot liquid erupts through weak points in the moon's crust.

The internal heating action is the result of a squeeze play put into motion by the gravity of Jupiter and the gravity of the three other large moons. Like the Earth's Moon, one side of Io permanently faces the giant planet. Jupiter's mighty gravity causes this side to bulge out toward Jupiter. But Io is also affected by the gravity of the other big moons. They cause Io's orbit to change shape. This, in turn, means that Jupiter's gravitational pull on Io varies. The bulge on the side facing Jupiter rises and falls. That action heats up Io's interior enough to melt it completely. Io's molten interior then punches through the cool crust, and a hot broth of sulfur gushes out.

Io's surface displays the same colors that sulfur has at different temperatures. The active volcanoes are brown, while the cooler plains are orange, yellow and white.

The Voyager close-up pictures of Io showed very few craters. Craters are caused by the explosive impacts of comets and asteroids smashing into the surface. The craters on our Moon have remained largely unchanged for three to four billion years. But if the craters are on a geologically active world, they will be erased, as has happened on Earth.

On Io, there are virtually no craters because the surface is constantly being blanketed by erupting material from volcanoes. Io literally turns itself inside out every one to two billion years.

LEFT **The scarred volcano-pocked face of Io was first seen by Voyager 1 in 1979. Until then, we knew almost nothing about this strange world. In 1995, the Galileo spacecraft, shown here, sailed past Io on its way into orbit around Jupiter.**

ABOVE **Jupiter's moon Io is the most volcanically active body in the solar system. In this illustration, artist David Egge keeps us at a safe distance from a volcano that is flooding the frozen landscape with hundreds of tons of yellow sulfur each second.**

EUROPA

Jupiter's moon Europa just might be the best place in the solar system to look for extraterrestrial life—not on its surface, but deep inside.

From a distance, Europa appears as smooth and unmarked as a ping-pong ball. At close range, though, the surface of this world is covered with thin lines that look like cracks in an eggshell. And that, in fact, is what they are—cracks. But in this case, the shell is made of ice. The surface of Europa is almost pure water ice, somewhere between 10 and 100 kilometers (6-60 mi) thick. That icy skin forms a casing around a subsurface ocean of liquid water and ice slush. The ocean, in turn, covers a rocky core.

The same kind of tidal heating that causes Io to spurt volcanoes (see page 26) keeps Europa's interior warm as well. But the effect is not as strong, and when the twin Voyager spacecraft passed Europa in 1979, they saw no steamy vents.

However, Europa's interior is warm enough to keep the subsurface ocean from freezing. The surface ice crust probably has not melted for billions of years, but it appears to crack from time to time, perhaps flooding and resurfacing like a skating rink every few million years. That could happen when the moon gets whacked by a comet or an asteroid.

Soon after Europa's real nature was revealed by the Voyagers, researchers advanced the startling idea that the subsurface ocean on Europa could harbor life. They realized that Europa had probably not always been encased in ice. When the planets first formed, Jupiter was much hotter than it is today. As Jupiter compressed under its own weight, tremendous amounts of heat were released. For a few million years, the huge planet must have been like a miniature sun, shedding enough heat that Europa's surface might have been liquid water rather than solid ice.

ABOVE The Sun slips behind Jupiter in this hypothetical scene from the icy surface of the giant planet's large moon Europa. RIGHT The surface of Europa was seen close up for the first time in 1979, when Voyager 2 hurtled by on its way to Saturn. The smooth surface of this virtually craterless ice world is broken by hundreds of cracks. Some of them are up to 20 kilometers (12 mi) wide, but all are filled in with ice. Below the global ice surface—possibly as much as 50 kilometers (30 mi) down—is an ocean of water that is kept from freezing by heat from Europa's core.

Some researchers speculate that during this warm phase, 4.5 billion years ago, life may have evolved on Europa, as it did in the ancient oceans on Earth. If it did, it might still be present in Europa's subsurface ocean.

Europa may therefore be the best prospect in the solar system for life—possibly better even than Mars, which no longer has any liquid water, the crucial ingredient for life as we know it.

Now, this is only a theory. But it is supported by the existence on Earth of a form of life that dwells in total blackness at the bottom of the ocean. Resembling a cross between weeds and worms, these creatures, called tubeworms, live around warm-water vents on the Earth's deep-ocean floor. The source of the warm water is seawater that seeps down through porous rock on the ocean floor, becomes

heated and saturated with minerals and hydrogen sulfide gas from exposure to hot rock below, then fountains up through chimneylike vents.

Hydrogen sulfide is poisonous to ordinary forms of life. But certain kinds of bacteria can thrive on this gas when it is combined with nitrogen, phosphorus and other nutrients in the hot water. In a bizarre mutually beneficial relationship, the tubeworms absorb the hydrogen sulfide and nutrients for the bacteria and the bacteria, in turn, release sugars to feed the tubeworms. Until this discovery of living creatures thriving in total blackness, sunlight had always been considered the essential energy source for all life forms on Earth.

On Europa, the heat from the rocky core, which prevents the ice-topped ocean from freezing solid, could create the same type of ocean-bottom vents. If life evolved in the primal open oceans that might have existed on Europa when Jupiter was young, it may still be there, feeding in darkness around subterranean ocean-bottom vents.

GANYMEDE AND CALLISTO

Jupiter's two largest moons are the same size as the planet Mercury, but their cratered surfaces are more ice than rock.

On the night of January 7, 1610, Italian scientist Galileo Galilei examined the planet Jupiter with a telescope he had built himself. Close to Jupiter, he noticed three small starlike objects. A week later, he spotted a fourth. Galileo soon became convinced that the four objects were circling Jupiter, just as the Moon orbits our planet. These four large moons—Io, Europa, Ganymede and Callisto—are still sometimes called the Galilean satellites, in honor of their discoverer.

In 1979, both Voyager 1 and Voyager 2 made close flybys of Jupiter and its moons. The powerful cameras on the twin spacecraft revealed incredible detail—and plenty of surprises.

Callisto, the outermost of the Galilean moons, is plastered with craters, much like the Earth's Moon and the planet Mercury. However, close inspection of the Voyager images revealed that Callisto is the most densely cratered body in the solar system. Early in the solar system's history, billions of planetesimals crashed into the planets and their moons. But the region near Jupiter was especially heavily hammered, because Jupiter's enormous gravity acted like a magnet, drawing in planetesimals from surrounding space. This still happens today. In 1994, a comet captured by the immense planet's gravity plunged into Jupiter's clouds (see page 50).

A walk on Callisto would be like an exploration of our own Moon. The two bodies have a similar surface gravity, and neither has an atmosphere. But the similarity would end with the collection of surface material. Bits of rubble and chunks of rock gathered from Callisto would melt at room temperature. Callisto's "rock" is actually mostly water ice, with lunarlike dirt mixed in. At Callisto's surface temperature of minus 145 degrees C (–230°F), water ice is as hard as stone. Deep in Callisto's interior, a rocky core probably exists, but the bulk of the world beyond the core region is largely water ice.

Ganymede, the largest of the Galilean satellites and the largest moon in the solar system, is slightly bigger than the planet Mercury. Like Callisto, it is a cratered world with a mainly water-ice surface darkened by dirt. Some of the major crater impacts have punched through the light brown soil-ice, exposing brighter, cleaner ice underneath. The ejected icy material is splattered around these craters in what astronomers call ejecta blankets.

But there is more than craters on the surface of Ganymede. Large areas of ribbed and ridged icy material are spread like continents over the satellite. These areas have fewer craters than other sectors and are clearly younger. Internal heat from Ganymede produced these features, which in some ways resemble glacial flows on Earth.

The remaining members of Jupiter's family of 16 known satellites are much smaller than the Galilean moons. The largest, Amalthea, is a potato-shaped chunk of rock less than one-tenth the diameter of Europa, the smallest of the big moons. The dozen small moons divide into three distinct families of four moons each:

- **Inner Group** Metis and Adrastea orbit near the outer edge of Jupiter's tenuous ring. Amalthea and Thebe are located halfway from the ring edge to Io's orbit.
- **Middle Group** Leda, Himalia, Lysithea and Elara orbit Jupiter in a belt six times farther out than Callisto's orbit.
- **Outer Group** Ananke, Carme, Sinope and Pasiphae occupy another belt twice as remote as the middle group. These moons are very small and are almost certainly captured asteroids.

LEFT **Because the surface of Ganymede is a mixture of pure water ice and material similar to sand or garden dirt, it is various shades of brown and gray. The white patches are scars caused by comets that long ago smashed into this moon.**

RIGHT **Jupiter and its largest moons. This mosaic of Voyager images shows Callisto at lower right, Ganymede at lower left, Europa at center and Io at upper left. Jupiter shows its famous Great Red Spot.**

SATURN

Saturn is adorned by a magnificent ring system that makes the planet one of the most majestic sights in the heavens.

Imagine a gigantic sphere more than nine times the Earth's diameter. The sphere consists almost entirely of gas and clouds: hydrogen, helium, methane, ammonia. Now imagine that this colossal object is so light, it could float in water—if a large enough ocean could be found to launch it. Finally, imagine the great world encircled by a trillion tiny moons spun into a thin disk. What do we have?

Saturn. The second largest planet. The least dense planet. And the most beautiful planet.

Viewed through a telescope, Saturn is truly one of the great spectacles of nature. Even a small telescope magnifying only 30 times will show the planet's rings, which appear like handles on either side of the planet. In a large telescope, the view is awesome. You can see the main division in the ring system, called Cassini's division, as well as pale cloud belts on Saturn itself.

Saturn's ring system is huge—from the inner edge to the outer edge, it is more than 10 times the width of North America. But the whole affair is no thicker than a 50-story skyscraper. As far as we know, Saturn's rings are the thinnest structure in nature. If the ring system were the width of a football field, it would be thinner than the paper this page is printed on.

The rings are made up of swarms of icy particles, ranging in size from tiny crystals like those in an ice fog to a few giant chunks as big as small mountains. If a particle the size of a baseball from Saturn's rings were returned to Earth and placed on a lab table, it would turn into a puddle of mud in a few minutes. Ring particles are mainly ice with dirt mixed in.

An exploration of the rings by a spacesuited astronaut outfitted with a propulsion backpack for maneuvering would be a spectacular adventure. The astronaut could float along with the ring pieces, touching, turning and holding them.

For every house-sized ring boulder, there are a million the size of a baseball and trillions the size of a grain of sand. In denser sections of the rings, the baseball-sized particles are likely separated by just a few meters, while the relatively rare house-sized ones would be several kilometers apart.

Research in recent years points more and more toward the idea that Saturn's rings are a temporary rather than a permanent structure. Saturn was not born with this adornment.

The rings may have been created when a comet smashed into one of Saturn's satellites and much of the debris remained in orbit around Saturn. An alternative theory is that two of Saturn's smaller moons collided and broke up and the pieces continued to smash into each other until the flat arrangement we see today was formed. Slowly, over time, the ring material will fall into Saturn, but it may take billions of years.

Saturn is nine times farther from the Sun than Earth is, which means the rings and moons of Saturn are cold. The average temperature on the surface of a ring particle or one of Saturn's icy moons is minus 200 degrees C (–330°F). At that temperature, water ice is as hard as rock. Most of Saturn's moons have been frozen rock-hard since the birth of the solar system, 4.7 billion years ago.

ABOVE **Saturn would fill the sky when viewed from the surface of its moon Enceladus. Because Enceladus orbits almost exactly above Saturn's equator, the rings are seen edge-on and look like a thin wire slicing across the face of the giant planet. But the shadow of the rings reveals sunlight pouring through a multitude of gaps.**

RIGHT **Saturn shows off its spectacular rings in this 1994 Hubble Space Telescope view. Like Jupiter, Saturn is a huge world made up almost entirely of gases. Its visible "surface" is a cloud layer with much less color and detail than Jupiter's cloud belts.**

TITAN

Saturn's largest moon, Titan, is the only world in the solar system that has an atmosphere similar in density and composition to the Earth's.

Titan is the ninth largest world in the solar system. Among moons, just Jupiter's Ganymede is slightly bigger. And both Ganymede and Titan are larger than the planet Mercury. But Titan stands alone as the only moon with a substantial atmosphere.

Titan is the only place in the solar system which has an atmosphere even remotely similar in composition and density to that of the Earth. The Earth's atmosphere is 78 percent nitrogen and 21 percent oxygen, while Titan's atmosphere is about 95 percent nitrogen and 5 percent methane, with traces of other gases. At the surface, the atmospheric pressure on Titan is 1.5 times what it is on our planet. No other known world comes close to matching Earth in this regard. The big difference between the two, though, is temperature.

The Earth's average surface temperature is a few degrees above the freezing point of water, but Titan's is minus 180 degrees C (–292°F), barely above the freezing point of methane. Just as water on Earth exists in solid, liquid and gaseous states, so methane on Titan exists as a solid, a liquid and a gas. Methane is commonly called natural gas. On Earth, it is extracted from the ground and shipped by pipeline to homes and businesses to be used as fuel.

When viewed from afar, Titan looks entirely featureless, like Venus. It is a world cloaked in a yellowish brown smog of methane, ethane and acetylene—compounds often found at oil refineries here on Earth. On the surface, there must be glaciers of methane ice, lakes of liquid methane and, in the lower levels of the nitrogen atmosphere, methane vapor that could fall as methane rain.

Humans protected by well-insulated heated spacesuits would have no problem exploring Titan. The ground would be a combination of methane ice and water ice, probably with a covering of a slushy mixture of methane rain and snow. When it snows on Titan, the methane flakes would seem to be falling in slow motion, since gravity is only one-fifth that on Earth, about the same as on the surface of our Moon.

The combination of low gravity and dense atmosphere on Titan makes it the best place in the solar system for flying. Aircraft with puny motors would easily stay aloft. Gliders would soar almost endlessly. Even human-powered flight with wings attached to our arms might be possible if cumbersome spacesuits are someday overcome. There would be a few hindrances, such as the need for

LEFT **Saturn, viewed through the smog and haze well above Titan's surface.**
CENTER **The surface of Titan is believed to be an unearthly panorama of glaciers made of methane and water ice. There would also be ponds and lakes of methane —liquid natural gas—that future explorers would have to negotiate.**
RIGHT **To an approaching space probe, Titan looks as featureless as a billiard ball. Clouds and smog completely envelop this intriguing world in an orangish blanket.**

floodlights, flashlights and helmet lamps to penetrate the gloom: by the time light from the Sun—only 1 percent its intensity at Earth—has filtered through the dense smog and clouds of Titan's atmosphere, it illuminates the surface with a glow barely brighter than our full Moon.

Titan's sky would appear a dull brownish color overall, without any definite indication of clouds. A day on Titan would be like a typical gray winter day on Earth, but much dimmer. And on Titan, a day is 16 Earth days long—8 days of light and 8 days of pitch-darkness.

An expedition by human explorers on the surface of Titan would resemble a similar exploration of Antarctica: intense cold, seemingly endless stretches of ice and snow, relatively little change in weather. There are probably areas where the surface ice is melted, perhaps mushy sections, maybe ponds or lakes of liquid methane. An interesting property of a methane lake is that it never freezes over. Methane ice sinks, so the ice builds up from the bottom. On Titan, boats may be a 21st-century mode of transport.

OTHER MOONS OF SATURN

Saturn rules a family of 18 moons that range in size from flying mountains to worlds of rock-hard ice. There is still much to be learned about them.

In his classic science fiction novel *2001: A Space Odyssey,* Arthur C. Clarke placed an alien civilization's Star Gate on Saturn's third largest moon, Iapetus (pronounced eye-A-pet-us). The Star Gate, an immense black slab, was a passageway to far distant realms of the universe. Clarke chose Iapetus because it is unique in the solar system: one side of this moon is mysteriously 10 times brighter than the other. In Clarke's novel, the Star Gate stands exactly in the middle of the bright side.

In 1980, Iapetus was photographed by Voyager 1. Although not very detailed, the pictures did reveal that the bright side of Iapetus is typical of most moons in the solar system. It is cratered and looks just like the neighboring moon Dione. (Dione, slightly smaller than Iapetus, is shown in the foreground on the facing page.) The dark side of Iapetus, however, is intriguing: It is virtually featureless and as black as charcoal dust. The Cassini space probe, expected to reach Saturn in 2004, should solve this enigma when it takes high-resolution shots of Iapetus.

Among the remainder of Saturn's family of moons, five rank as intermediate-sized worlds, ranging from one-half to one-tenth the diameter of the Earth's Moon. These are Rhea, Tethys, Dione, Enceladus and Mimas. All five, as well as the 11 smaller moons that orbit Saturn, are airless ice worlds.

Here, at Saturn's distance from the Sun, where the temperature never varies much from an average of minus 200 degrees C (–330°F), water ice is harder than cement. These moons are all heavily cratered worlds. One of them, Tethys, in addition to its craters, is scarred by a gigantic trench. The feature is 100 kilometers (60 mi) wide and 5 kilometers (3 mi) deep, and it extends halfway around the moon. Known as Ithaca Chasma, this huge ditch may have been caused by a collision with a smaller moon billions of years ago.

LEFT **The battered face of Saturn's moon Mimas is silent testimony to the bombardment that occurred throughout the solar system billions of years ago.**

RIGHT **Saturn and five of its moons. This mosaic of six images obtained by Voyager 1 in 1980 shows (from left to right) Rhea, Enceladus, Dione, Saturn, Tethys and Mimas. Because the original images were taken when Voyager was at a different distance from each object, these worlds are not shown in the correct relative sizes.**

URANUS

The seventh planet from the Sun has a weirdly tipped axis, a family of 15 moons and a system of skinny black rings.

Although four times the size of Earth, Uranus (pronounced YER-a-nus) is so remote that it appears as merely a pale greenish ball in the largest telescopes on Earth. Not even the planet's rotation period—its day—was known with any accuracy until the first spacecraft encounter by Voyager 2 in 1986. Voyager's cameras detected a few clouds in the planet's cloak of bluish green smog. Tracking these features revealed that the planet turns at different rates (14 to 17 hours) at different latitudes.

Nothing like the colorful cloud bands of Jupiter were seen by Voyager. On Uranus, a haze of methane ice crystals—a sort of ice fog—acts to block what lies beneath. The spacecraft captured images of some rare towering clouds, elevated by exceptionally energetic eddies from below. Occasionally, such clouds are also visible in shots taken by the Hubble Space Telescope.

At Uranus's core is a rocky structure about the size of Earth. This core is encased in a much greater mass of icy slush—mostly water mixed with liquid methane—thousands of kilometers deep. Overlying all that is an equally thick soupy atmosphere composed of seven-eighths hydrogen and one-eighth helium. What we see is the smoggy, generally blank top layer of the atmosphere.

Uranus's best-known feature is its oddly tipped rotation axis, which is only eight degrees from being horizontal. This flopped axis might have been caused by a collision with a body the size of Earth during the formation of the solar system. However it happened, the result is that the planet seems to roll around the Sun in its 84-year orbit, rather than spin in a near-vertical position as does Earth. Consequently, regions near one pole or the other are completely without sunlight for periods ranging up to 42 years (half a Uranian year).

Uranus's nine skinny rings are narrow and as black as soot, the opposite of Saturn's. Another difference between the rings of Uranus and those of Saturn (as well as those of Jupiter, which are entirely fine dust) is the particle size. When Voyager's radio signal was deliberately passed through Uranus's rings, analysis of the subsequent distortion indicated that they consist of icy boulders typically the size of a refrigerator.

ABOVE **Uranus floats in the cold blackness of the outer solar system in this hypothetical view from Ariel, one of its larger moons. Even at this close range, the bluish green haze that tops Uranus's atmosphere appears completely featureless.**

RIGHT **A view of Uranus's moon Ariel, taken by Voyager 2 from a distance of 130,000 kilometers (80,000 mi). Although its surface looks like rock, this world is almost entirely ice. The origin of the lengthy 25-kilometer-wide (15 mi) canyons is unknown, but astronomers suspect that sections of this moon melted three to four billion years ago. Ariel is about one-third the diameter of the Earth's Moon.**

We knew very little about the most distant of the giant gas planets until 1989, when the remarkable Voyager 2 space probe unveiled the great blue world.

Suppose you are at the controls of a spaceship traveling 10 times faster than any space probe ever built. That's about 400,000 kilometers per hour (250,000 mph). You decide to set a course for Neptune. After leaving Earth, you pass the Moon in less than an hour. After six days of travel, you reach Mars. But the voyage to Neptune is an excursion to the edge of the solar system. The trip to Neptune would take almost *17 months*!

Long before your arrival at Neptune, Earth would have faded to a faint, remote speck of light, lost from view in the glare around the Sun. As for the Sun itself: at Neptune, it shines with only 1/400 the brightness we experience on Earth. That is still brighter than 500 full Moons, so Neptune does not roam in complete darkness. But here at the rim of the solar system, it is frigid—only 40 Celsius degrees (70F°) above absolute zero.

At 30 times the Earth's distance from the Sun, Neptune is so remote and inconspicuous, it was not discovered until 1846, more than two centuries after the invention of the telescope. Over the next 140 years, astronomers learned that Neptune is a gas giant, basically the same as Jupiter, Saturn and Uranus. And they determined that Neptune and Uranus are about the same size—four times the Earth's diameter. Two of Neptune's moons were discovered, and Neptune was found to orbit the Sun once every 165 years. Until 1989, this meager list summed up our knowledge of the eighth planet. That was the year of the Voyager 2 space probe's spectacularly successful flyby of Neptune.

Voyager's discoveries began to pile up five months before it encountered Neptune on August 28, 1989, when the spacecraft's cameras picked up a dark patch bigger than Asia floating in Neptune's blue atmosphere. It was dubbed the Great Dark Spot. The huge feature proved to be a previously unknown atmospheric whirlpool something like Jupiter's Great Red Spot. Amazingly, Neptune's Dark Spot had exactly the same proportions and was located at the same latitude south of the equator as Jupiter's Red Spot. However, the Great Dark Spot is not a long-lasting feature like Jupiter's spot. By 1994, the Hubble Space Telescope could see no sign of it.

Why is Neptune blue? Methane clouds and haze in Neptune's upper atmosphere absorb red light and reflect blue light. Sunlight contains both colors, so an observer looking down on Neptune sees just the blue reflected light.

Methane is commonly known as natural gas. It liquefies at around minus 173 degrees C (–280°F) and freezes a few degrees below that. Neptune's upper atmosphere is near this temperature

LEFT **Three skinny rings surround Neptune, each made up of collections of rocky rubble ranging in size from pebbles to house-sized chunks. They may be the remains of small moons that collided and broke apart long ago.** RIGHT **Neptune's highest clouds are wispy streams of ammonia ice crystals that cast shadows on the blue ammonia mist below. The temperature at the level of the clouds is a frigid minus 175 degrees C (–283°F).** FAR RIGHT **Neptune, the fourth of the giant gas planets, is cloaked in a beautiful blue methane mist. Both images on this page were taken by Voyager 2 in August 1989, as it hurtled past the planet at 60,000 kilometers per hour (37,000 mph).**

and is made up mostly of tiny droplets of methane that form a blue mist. A few high cirrus-type clouds of white methane crystals stream above the main blue deck.

Below the visible "surface" of Neptune is a thick atmosphere of hydrogen and helium that, thousands of kilometers down, eventually merges with a warmer slush of hydrogen, helium and water ice sloshing around inside the planet. This core region rotates once in 16 hours, but the outer atmosphere spins more slowly and at different rates at various latitudes. The different rotation rates create friction, which produces heat inside the planet. The heat percolates to the surface and generates turbulence—Neptune's version of weather—which Voyager 2 saw as the Great Dark Spot and other features.

Although Neptune is almost identical in size to its neighbor Uranus, there is a difference in the amount of heat emerging from their interiors. Heat reaching

the surface from within Neptune exceeds the heat it receives from the Sun; with Uranus, the reverse seems to be true. Neptune's hotter interior produces the visible cloud features. Uranus's cooler interior does not generate as much turbulence at the surface cloud level. This seems to be why Uranus is not marked by any major belts or spots.

After the discovery of rings around Uranus in 1977 and Jupiter in 1979, an all-out search for Neptune's rings was made using large telescopes on Earth. This research showed that something was near the planet, but astronomers could not be sure exactly what it was. Finally, Voyager 2 solved the mystery.

Neptune sports five rings: three thread-thin ones and two wider ones which are so diffuse that Voyager 2 could barely detect them. Strangely, the three thin rings are thicker in some places than in others, which explains some puzzling observations from Earth. But what causes these clumps in the thin rings remains unknown.

TRITON

Neptune's largest moon was seen in incredible detail by Voyager 2 in 1989. Voyager pictures revealed a world of frozen nitrogen plains stained by amazing geysers.

Some people are way ahead of their time. One such visionary was American inventor Robert Goddard. From a farmer's field in 1926, he launched the first rocket with a liquid-fuel engine. Forty-three years later, the same basic type of rocket sent astronauts to the Moon. Although Goddard's experiments were ignored at the time, he is regarded today as the inventor of the space rocket.

There is a fascinating link between Goddard and Neptune's largest moon, Triton. A few months after he launched his first rocket, Goddard wrote an essay about travel to the planets and beyond. He predicted that voyages to the stars would be launched from a rocket base on Triton. He suggested that astronauts of the future would prepare spaceships for their long journeys at this remote outpost of the solar system.

We still don't know whether Goddard will be proved right about his Triton prediction, but we do know that he was afraid to show it to anyone, fearing that they would suspect he was completely out of his mind. He had his papers locked in a friend's safe so that no one could accidentally read them.

Robert Goddard died in 1945, long before Voyager 2, the first spacecraft to visit the Neptune system, reached its goal in August 1989. What would Goddard have thought if he had lived to see Voyager's close-up pictures of Triton?

Through Voyager's electronic cameras, we now have a spectacularly detailed portrait of an amazing world—the coldest place in the solar system. On Triton, the surface temperature seldom rises above minus 240 degrees C (−400°F). At that temperature, water ice is as hard as granite and liquid nitrogen acts like water, forming a subsurface layer something like the water table below the Earth's surface. Natural heat from inside Triton, caused by tidal action from nearby Neptune and by the decay of naturally radioactive elements, produces the subsurface nitrogen "groundwater."

Much of Triton is covered with brilliant blue-white nitrogen snow. The snow is thought to come from liquid and gaseous nitrogen that shoots into the sky from geysers erupting through cracks in the moon's outer crust. The liquid nitrogen freezes into crystals above Triton's surface, then falls to create white snowy plains. Frozen methane (solid natural gas) is also released during these eruptions, leaving purple and black stains on the ground beside the vents. Dozens of these stains are visible in the Voyager

ABOVE **Neptune's moon Triton was seen in detail for the first time, as depicted in this illustration, when Voyager 2 swept past it on its way out of the solar system. Dark streaks are downwind trails left by liquid-nitrogen geysers on this frigid world.**
RIGHT **Geysers on Triton, Neptune's largest moon, spew liquid nitrogen hundreds of meters skyward even though the temperature on this remote world is only 40 Celsius degrees (70F°) above absolute zero.**

images and are shown in the illustration on the facing page.

The dark stain marks are elongated by prevailing winds on Triton. Winds are not common on planetary satellites. Most, like our Moon, are airless, but Triton has a thin atmosphere of nitrogen and methane that is maintained by the geyser emissions. The atmosphere is clear and invisible except for a few wispy clouds.

Neptune may be off limits for human exploration, but that's not the case with Triton. Future Triton explorers will need heavily insulated heated boots, gloves and spacesuits to remain protected from the cold, but there is no reason why humans cannot visit Triton someday. The biggest problem will be getting there.

Future explorers may be able to confirm or refute the idea that Triton was once an independent planet made of ice and rock moving in its own orbit about the Sun. The theory suggests that Triton strayed too close to Neptune and was captured by the more massive planet's gravity about four billion years ago. For the next billion years, it swung in a long sausage-shaped orbit.

As it looped around this elongated orbit, Neptune's gravity raised giant tides that stretched and squeezed Triton. Astronomers calculate that this action generated enough internal heat to melt the entire world. It became a globe of molten rock and liquid water. Eventually, Triton's orbit became circular, and the surface froze. Any craters that Triton had were erased during this process, which is why so few are seen on this world today.

One of the main reasons astronomers support the capture theory is that Triton orbits its parent planet in a retrograde motion—that is, backwards compared with Neptune's rotation on its axis. Most moons orbit in the same direction as the planet's axis rotation. Astronomers think that these moons were formed at the same time as their parent planets. Backward moons are likely moons that have been captured. Jupiter has four of these, and Saturn has one—all of them almost certainly captured when they strayed too close to the big planets.

PLUTO

The ninth planet, at the rim of our solar system, is by far the smallest planet. It is also one of the most intriguing.

It's midsummer on the solar system's most remote planet. You might think it's the best time to visit Pluto, but don't switch on your spacesuit's air conditioner just yet. The temperature is minus 233 degrees C (–387°F), balmy only by Pluto's standards.

Summer on Pluto occurs when that planet is closest to the Sun, and it lasts about 50 years. It's also the only season when the planet has an atmosphere—a tenuous veil of nitrogen and methane gas created as a small amount of Pluto's icy surface is vaporized by the feeble heat from the remote Sun.

By the year 2020, the atmospheric gases will once again be frozen to the ground as the distant planet swings farther from the Sun and the century-long Plutonian winter approaches.

The sluggish cycle of seasons on Pluto is the result of its huge egg-shaped 248-year orbit, which carries the tiny planet from a near point of 30 times the Earth's distance from the Sun to a far point of 50 times the Earth-Sun distance. For 20 years of each orbit, Pluto is closer to the Sun than Neptune is and then ranks as the eighth planet. That's where it is now. Neptune regains eighth position in 1999. In the diagram on page 47, you can see how Pluto's orbit carries the planet from inside Neptune's orbit to almost twice Neptune's distance.

Although the orbits of Pluto and Neptune appear to cross, Pluto's orbit is angled so that it passes above Neptune's orbit, like a freeway interchange. The two orbital paths are actually millions of kilometers apart.

Furthermore, Pluto's orbital period is 1.5 times that of Neptune. This situation is called orbital resonance. It means that Neptune orbits the Sun three times in the same period it takes Pluto to orbit the Sun twice. When Pluto is at its closest point to the Sun during this celestial waltz, as it was in 1989, Neptune is always well away from the region where their orbits overlap. This keeps Neptune's gravitational influence on Pluto to a minimum. If Pluto were not in orbital resonance with Neptune, Neptune's gravity probably would have kicked Pluto out of its orbit entirely, possibly even out of the solar system.

At Pluto's minimum distance, the Sun appears no larger than the head of a pin held at arm's length. But even so, it is not perpetual night out here in the back row of the solar system. Sunlight is still strong enough to illuminate Pluto's surface more than 600 times as brightly as the full Moon does on Earth.

Pluto remains the only planet in our solar system not yet examined close up by spacecraft. There were preliminary plans to send the American Voyager 1 to Pluto after its swing by Saturn in 1980, but to do so would have meant sacrificing a close look at Titan, Saturn's largest moon. Since there was no guarantee that Voyager 1 would still be working during the Pluto encounter a decade later, mission planners went for Titan.

We now know for certain that Pluto is by far the smallest planet in our solar system. It is only 2,320 kilometers (1,440 mi) in diameter, less than one-fifth the width of Earth and less than half the size of Mercury, the next smallest planet. As small as Pluto is, though, it has a comparatively huge moon named Charon (pronounced CARE-on). Charon is 1,270 kilometers (790 mi) in diameter, slightly more than half Pluto's diameter. We could say that Pluto is the solar system's only double planet.

Though Pluto was discovered in 1930, Charon was not detected until 1978. This is because the two worlds are blurred into a single image when viewed through ground-based telescopes. The blurring is caused by ever present turbulence in the Earth's atmosphere. That's why the Hubble Space Telescope, orbiting high above the Earth's atmosphere, can capture images so much sharper than telescopes on the surface of Earth.

What we have been able to learn from the Hubble Space Telescope and other Earth-based telescopes indicates that Pluto is about the same size, mass and composition as Neptune's moon Triton. The interiors of both worlds are probably made up of a mixture of roughly equal amounts of ice and rock. Pluto's surface, like that of Triton, is mainly water ice with some nitrogen ice.

Since Pluto's orbit crosses the orbit of Neptune, some astronomers have suggested that Pluto may be an escaped moon of Neptune. But this idea is now considered highly improbable. Pluto and its moon likely formed in the region where we see them now.

LEFT **While Pluto is cruising the section of its orbit that carries it closest to the Sun, the Sun vaporizes a thin layer of surface ice, giving Pluto a tenuous atmosphere of nitrogen and methane. Sunlight here is 600 times stronger than full moonlight on Earth, but the temperature is a very cold minus 233 degrees C (–387°F).**

ABOVE **Pluto and its moon Charon, as seen by the Hubble Space Telescope. Charon is half the size of Pluto, which makes it by far the largest moon in the solar system relative to the size of its parent planet. Since no space probes have visited Pluto, this is the most detailed picture of the planet and its moon that exists.**

Pluto is the largest known member of the Kuiper belt, a region just beyond Neptune's orbit where millions of comets roam.

In 1929, Clyde Tombaugh, a 23-year-old amateur astronomer from Kansas, read about studies of the planet Mars under way at Lowell Observatory in Flagstaff, Arizona. Tombaugh, who had been watching Mars himself with a homemade telescope, mailed some of his sketches of the red planet to astronomer Vesto Slipher, the observatory director.

Slipher was so impressed by Tombaugh's letter and sketches that he offered him a job at the observatory. His task was to search for a planet beyond Neptune. A year later, Tombaugh became world-famous as the discoverer of the planet Pluto.

Tombaugh spent the next 13 years searching for a tenth planet beyond Pluto, but he found nothing. During the same period, astronomers were developing a theory about comets and speculated that there must be a comet storehouse beyond Pluto.

In 1950, Dutch astronomer Jan Oort proposed that billions of comets form a loose cloud beyond the planets, extending from 75 times Pluto's distance from the Sun to 1,500 times that distance. This region is now called the Oort cloud.

A year after Oort's proposal, American astronomer Gerard Kuiper (pronounced COY-per) suggested the existence of a second comet storehouse—a region something like the asteroid belt but located just beyond the orbit of Neptune. The proposed comet zone came to be known as the Kuiper belt. The inner section of the belt occupies the same region where Pluto orbits the Sun.

At the time of its discovery, Pluto was thought to be the size of Earth. Eventually, as telescopes improved, it became clear that Pluto is smaller than the Moon, made up mostly of ice, like a comet, and only 1/500 the Earth's mass. Astronomers were slow to make the connection between Pluto and the Kuiper belt, but we now know that Pluto is composed of basically the same kind of icy material found in comets. That makes it the Kuiper belt's largest member.

As of mid-1995, more than 20 objects, dubbed Kuiperoids, had been discovered within the Kuiper belt. The first one (other than Pluto) was spotted in 1992. About 200 kilometers (120 mi) in diameter, it orbits the Sun once in 291 years. The others are about the same size and have orbital periods ranging from 175 to 305 years. There may be millions of them larger than Deimos, the smallest moon in the solar system.

If Pluto were discovered today, it is unlikely that it would be called a planet. Rather, it would be listed as the largest Kuiperoid. At this point, however, astronomers are not about to change Pluto's status; it will continue to be regarded as the ninth planet.

Could there be a tenth planet? For many years, astronomers thought Uranus and Neptune deviated slightly from their predicted positions as they orbited the Sun. The gravitational influence of a tenth planet—Planet X—several times larger than Earth was suggested to explain the discrepancies. But search after search failed to reveal the hypothetical tenth planet.

More recently, astronomers re-examined the data which led to these predictions, concluding that all of the orbital "deviations" were in fact errors in measurement or calculations. There is nothing beyond Neptune, they now say, except the Kuiper belt and the Oort cloud.

LEFT **Pluto, as seen in this rendering from the surface of its moon Charon, is now known to be a comet, the largest member of the Kuiper belt. However, nearly three-quarters of a century after its discovery, astronomers are not about to demote Pluto from its planetary status.**

RIGHT **The Kuiper belt is located just beyond the orbit of Neptune, where icy debris left over from the formation of the giant planets collected into billions of huge snowballs—cometlike bodies dubbed Kuiperoids. Although the belt may look crowded, this diagram covers an enormous volume of space. The average distance between Kuiperoids is millions of kilometers. A second comet reservoir, called the Oort cloud, is much farther out.**

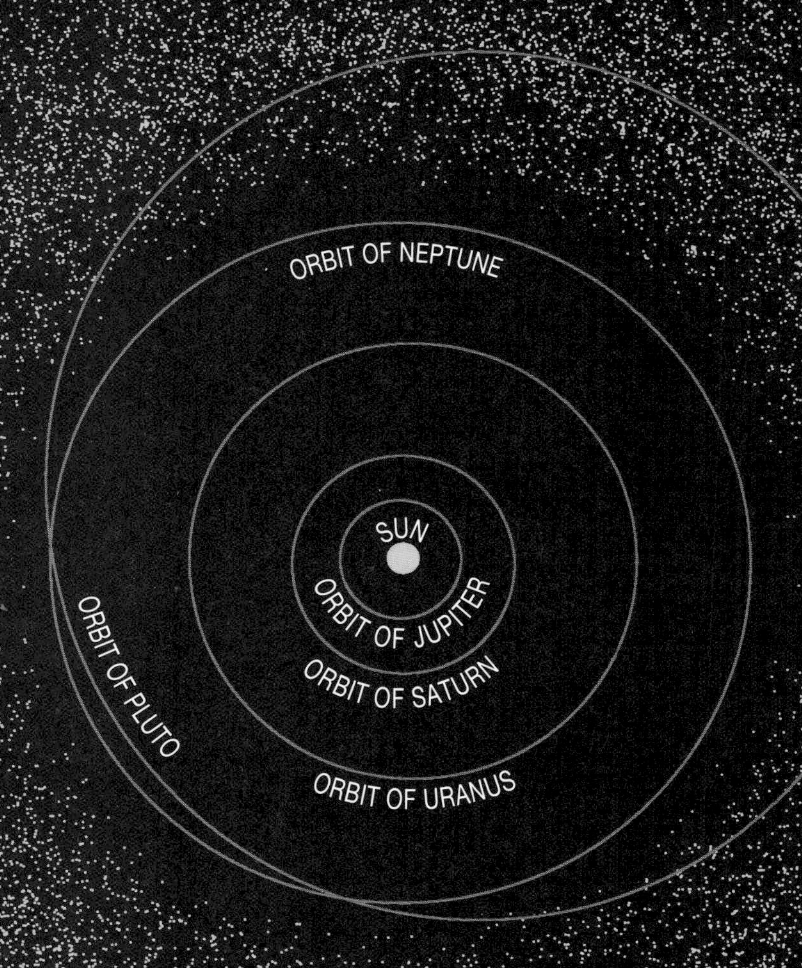
ORBIT OF NEPTUNE
SUN
ORBIT OF JUPITER
ORBIT OF SATURN
ORBIT OF PLUTO
ORBIT OF URANUS
KUIPER BELT

There are more than a trillion comets beyond the orbit of Neptune. A few eventually find their way into the inner solar system.

On their way to work just before sunrise on the morning of January 13, 1910, workers at the Transvaal Premier Diamond Mine in South Africa noticed what they described as "a star with a tail attached," near the eastern horizon. When they reached the mine, they reported the sighting to their foreman, who contacted the local observatory. The miners had discovered a rare "daylight" comet, one so dazzling that it could be seen beside the Sun in full daylight. Only three comets this bright have been observed during the 20th century.

Within two weeks, the Great January Comet, as it was called, had moved far enough from the Sun that it was visible around the world as a beautiful sight in the evening sky. It looked like a luminous feather hovering among the stars.

LEFT **Seen close up, the icy nucleus of a comet is surprisingly dark. That's because it is covered with layers of dust. When the comet nears the Sun, this dark material is heated by sunlight. Eventually, the ice beneath the dust vaporizes and erupts through the surface in "jets." Seen from afar, the vapors and dust from these jets form a cloud around the comet, which is pushed into a tail by the pressure of sunlight and by the solar wind.**

CENTER **View of a comet passing near the planet Mercury.**

RIGHT **Halley's Comet in 1986, when it was last visible to stargazers on Earth. To the naked eye, it appeared much dimmer than this telescopic view. Its next visit to the Earth's vicinity will be in the year 2061.**

Years later, many people thought they had seen the famous Halley's Comet, which was visible in the spring of 1910, though not as prominently.

Comets fascinated our ancestors more than any other celestial sight. What especially baffled them about these stellar vagabonds was the way a comet would appear as if from nowhere, march across the sky for a week or two sporting a pale glowing tail, then disappear. Not until the middle of the 20th century did we fully understand what these objects are.

A typical comet, like Halley's, is a flying mountain of ice, dust, dirt and gravelly rock. Trillions of comets orbit the Sun beyond Neptune, either in the Oort cloud or in the Kuiper belt. Up close, a comet would look like one of the small moons of Uranus or Neptune—basically a giant, dirty snowball.

Occasionally, a comet's orbit is affected by Neptune's gravity or by the gravity of the stars in our galaxy. The orbit then changes enough to carry the comet in toward the planets. This is what happened to Halley's Comet many thousands of years ago. It made a pass near Neptune and was catapulted by Neptune's gravity into a long, narrow orbit that carries it from Neptune's orbit inward to the orbit of Venus and back again over a period of 76 years.

Comets don't sprout tails unless their orbital paths carry them within the orbit of Mars, where the sunlight becomes strong enough to vaporize the ice. The vapor forms a huge cloud around the solid comet body. The cloud is pushed back into a tail by the pressure of sun-

light and by a flow of atomic particles from the Sun called the solar wind.

It may be hard to imagine how something as insubstantial as sunlight could have enough force to create a comet's tail. But in interplanetary space, there is no air for the comet to plow through. Tiny particles of dust ejected from the comet's solid nucleus are easily pushed around by the pressure of sunlight, even though it is weak. Since the dust particles are highly reflective (think of dust motes illuminated by a shaft of sunlight streaming into a dark room), the tail is the comet's most prominent feature.

Because a comet can zoom into the inner solar system from almost any direction, it appears without notice, providing great sport for amateur astronomers, who try to be the first to spot one. More ominously, comets can smash into planets and their moons. This doesn't happen often, but Jupiter got clobbered in 1994 (see page 50), and astronomers suspect that the object which struck Earth 65 million years ago and wiped out the dinosaurs was a comet.

COMET CRASH ON JUPITER

An event like the crash of Comet Shoemaker-Levy 9 into Jupiter in July 1994 occurs, on average, once in 1,000 years.

Toward the end of July 1994, thousands of professional and amateur astronomers around the world watched through their telescopes in amazement as huge black markings the size of Earth emerged in the southern-hemisphere cloud belts of the planet Jupiter.

The black spots—scars from the impact explosions of 21 mountain-sized pieces of Comet Shoemaker-Levy 9—became the most prominent features ever seen on any planet since the invention of the telescope. Some observers reported seeing them using telescopes as small as 60mm aperture at magnifications as low as 40 power.

The comet that produced the blemishes on Jupiter was discovered in 1993 during a routine search for comets and asteroids at Palomar Observatory in California. Like almost all comets, this one was named after its discoverers, Eugene and Carolyn Shoemaker and David Levy.

Calculations soon showed that the comet was orbiting Jupiter once every two years. On its previous pass in 1992, the comet came so close to the giant planet that it was split into 21 fragments by Jupiter's gravity. But the big surprise came when astronomers determined its future motion: Comet Shoemaker-Levy 9—all 21 pieces of it—was heading for impact on Jupiter in July 1994.

Before the collisions, no one knew what to expect when the comet fragments slammed into Jupiter. Nothing like it had ever been seen before: No substantial object had ever been observed crashing into a planet.

Each piece was hurtling at 60 kilometers per second (40 mi/sec) when it collided with Jupiter—so fast that in less than three seconds, it punched

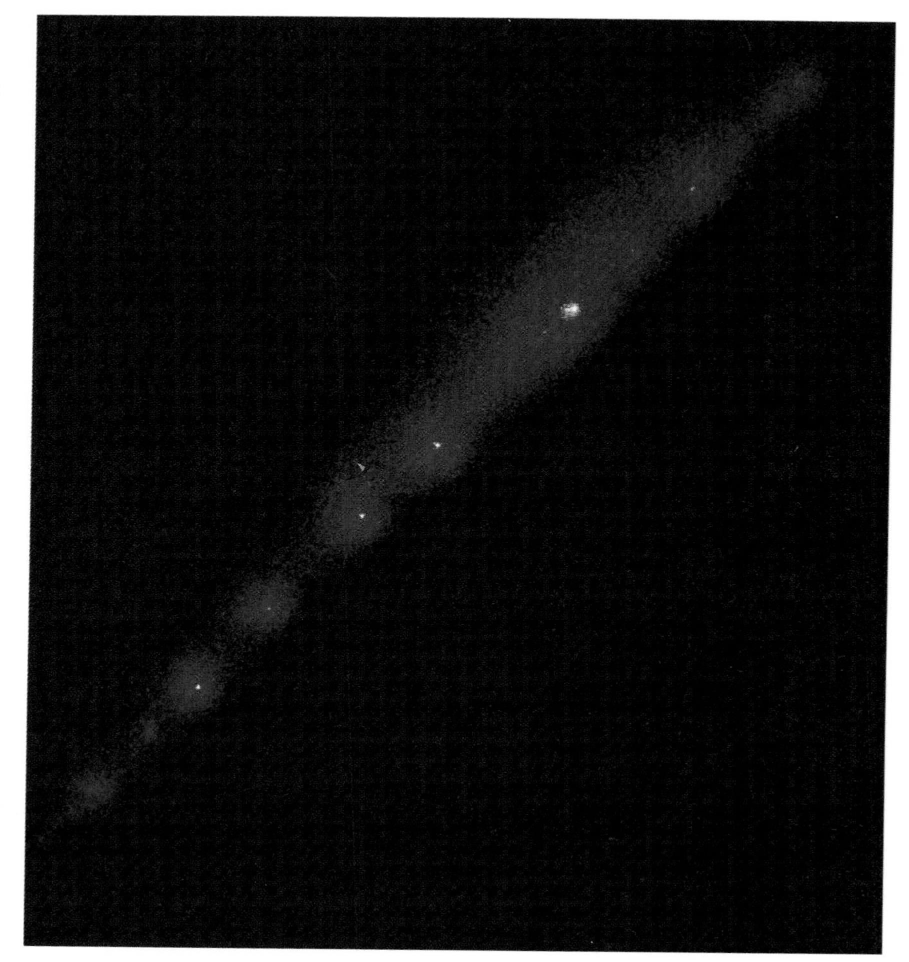

RIGHT **Like a string of pearls, the fragments of Comet Shoemaker-Levy 9 were strung out in a line as they headed to their doom in the clouds of Jupiter in mid-1994. The comet was broken into pieces by Jupiter's gravity when it passed close to the giant planet in July 1992.**

FAR RIGHT **Comet bruises on Jupiter, late July 1994, as revealed by the Hubble Space Telescope. The row of dark clouds on Jupiter shows where the fragments of Comet Shoemaker-Levy 9 collided with the planet over a period of one week.**

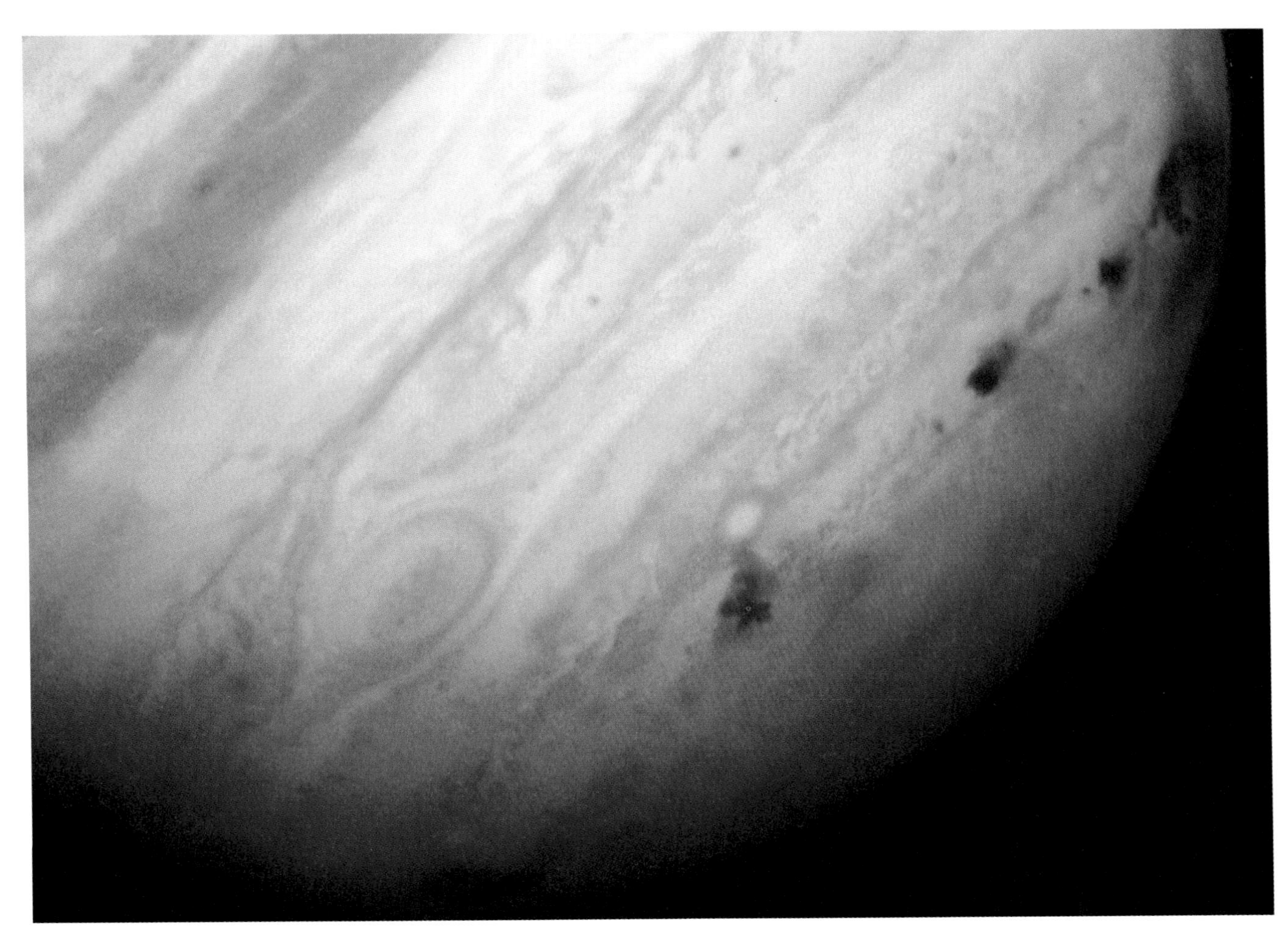

through Jupiter's upper atmosphere and plowed into the denser atmospheric layers below, where it was stopped like a bullet hitting a steel door. In a fraction of a second, the intruder's energy of motion was converted into heat, and the once frigid chunk of celestial ice vaporized and exploded in a colossal fireball.

The largest comet chunk—a piece of ice and rock possibly two or three kilometers (1-2 mi) in diameter—rammed into Jupiter's atmosphere and produced an impact explosion that released one million times the energy of a hydrogen bomb. It created a mushroom cloud larger than Asia that rose several thousand kilometers above Jupiter's cloud tops, then spread and cooled into a vast black patch the size of Earth. That patch remained visible for months, as it twisted in Jupiter's stratospheric winds.

The impacts from the comet fragments proved to be so spectacular and the media coverage so extensive that comet codiscoverer Eugene Shoemaker said he hadn't witnessed such excitement since the moon landings 25 years earlier.

BROWN DWARFS

Brown dwarf is the name used for objects more massive than Jupiter but less massive than small red dwarf stars.

The difference between a star and a planet is that the pressure at the core of a star is great enough to compress atoms of hydrogen into helium. This process, called fusion, releases colossal amounts of energy and is the power behind the light and heat of all Sunlike stars.

Any object with less than 80 times Jupiter's mass is not a true star, because it does not have sufficient core pressure to start hydrogen-fusion reactions. Objects with masses ranging from 10 to 80 times the mass of Jupiter could be called super-Jupiters. But astronomers have chosen to name them *brown dwarfs*. (Since the least massive stars are called red dwarfs, because of their color and small size, astronomers thought the name brown dwarfs was appropriate for even dimmer objects.)

Brown dwarfs fill the size gap between giant planets and bantamweight stars. But close up, they would probably look more like gas giant planets than stars. With surface temperatures of 1,000 to 1,500 degrees C (1,800-2,700°F), a brown dwarf might appear like a glowing ball of liquid, with ponderously churning currents of gas rising and subsiding in a constant play of colors and patterns.

A brown dwarf's heat comes not from hydrogen fusion but from gravity squeezing the object and compressing it. A compression of only a few millimeters per year would create enough heat to keep a brown dwarf's surface hot for billions of years. The compression also limits their size: all brown dwarfs would be roughly the same size as Jupiter.

Some astronomers suspect that brown dwarfs are at least as common as red dwarf stars, and there are probably 100 billion red dwarfs in our galaxy. Brown dwarfs have been predicted to be 10 times as abundant, but that's just a theoretical estimate. The fact is, the low surface temperature of brown dwarfs compared with true stars makes them extremely dim and difficult to detect. Despite decades of searching using many different and ingenious techniques, not a single brown dwarf has ever been positively identified. And this fact is beginning to worry astronomers.

If there are as many brown dwarfs as red dwarfs—or more, as a few theories suggest—some of them, it can be argued, should have been discovered by now. Researchers thought brown dwarfs would show up in images made with infrared cameras aboard scientific satellites that are sensitive to low-temperature objects. None did. They also looked for the telltale wobbling of nearby stars caused by the gravitational effects of a brown dwarf in orbit about the star. The results have not been convincing.

But there is one experiment that may

actually have detected a few brown dwarfs. The experiment is based on Albert Einstein's 1932 prediction that if two stars are seen exactly lined up, the more distant one would brighten because its light would be gravitationally lensed, or focused, by the nearer star, as if by a magnifying glass. In 1994, the telescope used for this experiment detected three stars brightening in exactly the way theorists say they would if brown dwarfs were acting as gravitational lenses.

Still, this is circumstantial evidence, not proof. As the mystery of the elusive brown dwarfs deepens, astronomers are wondering what's wrong with their theories. Most theories which predict how stars and planets are born suggest that there should be an abundance of brown dwarfs—possibly one trillion in our galaxy alone. Yet after decades of searching, we can't say positively that even one has been found.

LEFT A brown dwarf about 10 times the mass of Jupiter may resemble the giant planets in our solar system, as shown in this rendering. It could easily have a moon the size of Earth. Other hypothetical moons are shown in the foreground.

RIGHT Brown dwarfs are the missing link between giant planets like Jupiter and low-mass stars called red dwarfs. Although no one knows for certain what a brown dwarf would look like close up, this artist's conception by David Egge shows a colorful atmosphere swirled by heat welling up from a hot interior.

OTHER PLANETARY SYSTEMS

For decades, we have thought that many stars probably have planets orbiting around them, just as our Sun does. But that may not be true.

There are more stars in the universe than there are grains of sand on all the beaches on planet Earth. That's roughly one billion times one trillion stars—a truly colossal number (1 followed by 19 zeroes).

Most stars are fundamentally like our Sun. They release heat, light and energy in basically the same way the Sun does, and they were formed in clouds of gas and dust, again just like our Sun. It therefore seems reasonable to assume that there could be families of planets orbiting around many stars, just as the planets in our solar system orbit around the Sun. So far, though, this is just a theory. Despite 40 years of searching, researchers have been unable to confirm that another planetary system like our own exists.

Yet you'd think it would be easy to spot something the size of Jupiter orbiting another star. And, in fact, as long ago as 1964, astronomers believed they had found just such a planet orbiting a star known as Barnard's Star. They didn't actually see it, just evidence of it—or so they thought. Photographs taken of Barnard's Star as it moved through the galaxy appeared to show a wobbling motion caused by the unseen planet's gravity pulling on the star and causing it to deviate slightly from its straight-line movement through space. But this and several more recent reports of stars wobbling in their paths because of supposed Jupiter-sized planets in orbit about them have not been confirmed.

The disappointing news is that no planet has been found orbiting an ordinary star—one like the Sun. However, objects roughly the mass of Earth have been detected orbiting around two pulsars (see page 56).

If you have seen the movie *Star Wars*, you may remember a scene on Luke Skywalker's home planet showing a sunset with two suns hovering above the horizon. Is this possible, or is it pure fiction?

Planets of double stars have been a favorite science fiction setting for most of the 20th century. It's fun to think about scenes with yellow and red suns or blue and orange ones. As for the facts, astronomers know that about half the stars in the sky are members of double- or multiple-star systems.

Let's suppose a hypothetical planet is in orbit around a close pair of stars at the distance that Earth orbits the Sun.

ABOVE **Sunrise on a hypothetical planet with two suns. Although astronomers have no solid evidence that a planet like this exists, there is no reason why some planets could not have two suns instead of one. Many stars are double systems, and some are triple or quadruple. There are an almost infinite number of possibilities.**
RIGHT **Telescopes have detected disks of gas and dust swirling around some stars similar to the Sun. Astronomers think that planets may be forming from such material, as shown in this illustration, perhaps the way the planets in our solar system were created five billion years ago. The youthful planet at right shows the molten scars caused by asteroids and other debris crashing onto its surface.**

Astronomers tell us that such an orbit for a planet is possible, but over long periods of time, the orbit would become unstable. This is because the gravitational influences of two stars swinging around each other change the shape of the planet's orbit. The effect would be very slight over short periods of time, such as centuries. But during the billions of years a planet like Earth has been in existence, such an orbit would have changed so drastically that the planet would probably have been flung into deep space.

Now let's imagine a different scenario. Suppose that in our own solar system, the planet Jupiter, instead of being a gas giant planet, were a small star. If we were to make that star 100 times more massive than Jupiter, it would be a red dwarf star. The red dwarf could have its own family of planets, like an expanded version of Jupiter's present satellite system. Those planets could orbit the red dwarf star without any problem, while another family of planets orbited around the main star.

The difference, though, is that when an object as massive as a red dwarf star is sitting where Jupiter is, its gravitational influence affects the things around it. For instance, it would severely disrupt the orbits of the planets nearest it, like Mars and Saturn. But planets farther away, such as Mercury, Venus and Earth, would remain fairly stable.

PULSAR PLANETS

The first planets found beyond our solar system orbit a star that astronomers thought could not possibly have planets.

Thirteen hundred light-years in the direction of the constellation Virgo lies a tiny star not much larger than Mount Everest. The star has no name, just a catalog number: PSR1257+12. It is invisible except to radio telescopes that detect extremely precise pulses from the star as it spins 162 times per second and flashes radiation like a lighthouse beam.

Thousands of these twirling mini-stars have been discovered. Called pulsars, they are among the most bizarre members of the celestial zoo. Besides being small and whirling rapidly, pulsars are the densest objects known. A thimbleful of material from a pulsar, if brought to Earth, would weigh as much as a million railroad locomotives.

But there is something even more unusual about PSR1257+12: It has companions. Orbiting around the tiny pulsar —just as the planets wheel around our Sun—are two planets approximately the size of Earth.

Detecting the existence of planets orbiting a star 1,300 light-years away—especially such a small star—seems impossible. Yet the incredibly rapid spin rate of the pulsar means that it sends out signals—pulses—with a precision which rivals that of an atomic clock.

The discovery of the pulsar planets happened during routine pulsar monitoring with radio telescopes. Astronomers were measuring the precise arrival times of the pulses. But in the case of pulsar PSR1257+12, there was something puzzling about the results. Further examination of the pulse timings uncovered a strange cycle: the pulses arrived earlier, then later than expected over a period of months. Careful analysis of this cycle revealed the existence of the planets.

The astronomers realized that the gravity of the planets must be swinging the pulsar toward and away from Earth by just a few thousand kilometers. When the pulsar moves away, we receive the pulses slightly late, and when the pulsar moves toward us, we receive the pulses earlier. The difference is very slight, but because the pulsar's spin is so regular, the discrepancy is detectable.

In 1992, radio astronomers determined that the two planets orbit the pulsar once every 67 and 98 days at approximately the same distances from the pulsar as Mercury and Venus are from the Sun in our own solar system. They are, respectively, 3.4 and 2.8 times the Earth's mass. There are indications that a third planet, about the size and mass of the Moon, orbits the pulsar at roughly the Earth's distance from the Sun.

But conditions on these planets are completely different from those in our own solar system. Intense radiation sprayed out from the pulsar's gyrating magnetic field fries the surfaces of the pulsar planets. They must be molten slag, with no possibility of life as we know it. Calculations suggest that the two larger planets are about 1.4 times the Earth's diameter.

Until this discovery, a pulsar was the last place anyone expected to find a planet. According to theory, pulsars are the collapsed superdense cores of stars that explode as supernovas. Theoretically, any planets that exist around a star which becomes a supernova would be either incinerated, flung into space or sent into unstable, elongated orbits.

Astronomers think that the pulsar's planets are not survivors but, rather, new worlds formed from debris left over after the supernova explosion that created the pulsar in the first place. The planets are probably the remains of a companion star similar to the Sun that once orbited the pulsar. As it spiraled closer to the pulsar, this star ultimately broke up, leaving a disk of material from which the pulsar's planets formed.

LEFT **Astronomers never expected to find planets around a pulsar. But reality proved to be as bizarre as science fiction. For one thing, the two larger pulsar planets—both about 1.4 times the Earth's diameter—are 1,000 times bigger than the star they orbit. They are probably dense rocky bodies enriched with heavy metals such as iron and nickel. This rendering shows the more distant of the two larger planets.**

ABOVE **The surface of the nearest planet to pulsar PSR1257+12 must be hotter than Venus or Mercury—somewhere between an oven and a blast furnace. The heat comes from lethal gamma rays and x-rays emitted by the pulsar's magnetic field. In this environment, a human would last about as long as someone who was standing close to the flame of a space shuttle engine during liftoff.**

1957 U.S.S.R. puts first artificial satellite, Sputnik 1, into orbit on October 4. The space age is born.

1958 First U.S. satellite, Explorer 1, is launched on January 31.

1959 U.S.S.R. probe Luna 1 is first spacecraft to leave Earth's gravity.

1959 Luna 3 circles Moon and transmits first images of Moon's far side. (U.S.S.R.)

1961 Yuri Gagarin, first human in space, makes single orbit of Earth on April 12 in Vostok 1. Lands safely in U.S.S.R.

1962 John Glenn, first American in orbit, circles Earth three times.

1962 Mariner 2 passes Venus; first successful flyby of any planet. (U.S.A.)

1965 Mariner 4 makes first successful flyby of Mars. Sends back fuzzy pictures showing craters. (U.S.A.)

1966 Luna 9 sends back first pictures from surface of Moon. (U.S.S.R.)

1966 Venera 3 spacecraft makes first entry into atmosphere of another planet as it descends by parachute into dense clouds of Venus. Does not survive to surface. (U.S.S.R.)

1968 Release of landmark movie *2001: A Space Odyssey* heightens drama surrounding upcoming Apollo Moon flights and romances new generation of space-travel enthusiasts.

1968 Three astronauts aboard Apollo 8 are first humans to orbit Moon. Mission: to test Moon-flight hardware. (U.S.A.)

1969 July 20. Apollo 11 astronauts Neil Armstrong and Buzz Aldrin walk on Moon. More than one billion people watch the event live on television. (U.S.A.)

1970 Venera 7 lands on Venus and transmits data from surface. First successful landing on another planet. (U.S.S.R.)

1970 Wheeled moon rover aboard Lunokhod 1 is first mobile device on another world. (U.S.S.R.)

1971 Mars 3 lands on red planet but returns only 20 seconds of data. (U.S.S.R.)

1971 Mariner 9 orbits Mars, becoming first device to orbit another planet. Maps most of planet over next year. (U.S.A.)

1972 Apollo 17 mission ends the U.S. Moon-landing program. In six landings (1969-72), 12 astronauts explored lunar surface.

1973 Pioneer 10 becomes first probe to fly by Jupiter. (U.S.A.)

1974 Mariner 10, first spacecraft to reach Mercury, photographs planet during flyby. (U.S.A.)

1975 Venera 9 transmits first pictures from surface of another planet—Venus. (U.S.S.R.)

1976 First pictures from surface of Mars transmitted by Vikings 1 and 2. (U.S.A.)

1979 Pioneer 11 becomes first spacecraft to reach Saturn. (U.S.A.)

1979 Voyagers 1 and 2 fly by Jupiter and gather first detailed images of giant planet. (U.S.A.)

1980 First detailed pictures of Saturn—from Voyager 1 flyby. For first time, spacecraft photographs of a planet other than Earth grace covers of *Time*, *Newsweek* and other mainstream publications.

1983 Pioneer 10 reaches a point farther from Sun than the orbit of most distant planet (Pluto) and, in doing so, becomes first human artifact to leave solar system. (U.S.A.)

1983 Venera 15 orbits Venus and maps most of planet in low-resolution by radar. (U.S.S.R.)

1985 Vega 1 passes Venus and drops first balloon probe into another planet's atmosphere. (U.S.S.R.)

1986 Voyager 2 makes first flyby of Uranus and sends back excellent-quality pictures. (U.S.A.)

1986 European and Soviet spacecraft fly past Halley's Comet and gather first high-resolution images of a comet nucleus.

1989 Voyager 2 makes first flyby of Neptune and sends back excellent-quality pictures. (U.S.A.)

1990-93 U.S. Magellan spacecraft maps Venus in high resolution from orbit by radar.

1991 On its way to Jupiter, Galileo probe takes first close-up photographs of asteroid. (U.S.A.)

1994 Images of asteroid Ida taken by Galileo probe in 1993 reveal first moon of an asteroid. (U.S.A.)

1994 Refurbished Hubble Space Telescope allows regular high-resolution imaging of the planets to begin.

1995 Galileo's atmospheric probe will plunge into Jupiter's clouds and become first device to explore atmosphere of a giant planet.

Object	Diameter (Earth=1)	Diameter (kilometers)	Mass (Earth=1)
Sun	109.1	1,392,000	332,946
Mercury	0.38	4,879	0.06
Venus	0.95	12,104	0.82
Earth	1.00	12,756	1.00
Mars	0.53	6,794	0.11
Jupiter	11.19	142,980	317.8
Saturn	9.41	120,540	95.2
Uranus	4.01	51,120	14.5
Neptune	3.88	49,530	17.2
Pluto	0.19	2,320	0.002

Object	Distance from Sun (Earth's distance=1)	Length of Year (one orbit of Sun)	Length of Day (sunrise to sunrise)
Mercury	0.39	88.0 days	176 days
Venus	0.72	224.7 days	117 days
Earth	1.0	365.3 days	24 hours
Mars	1.5	687.0 days	24h 39m
Jupiter	5.2	11.86 years	9h 50m
Saturn	9.5	29.46 years	10h 39m
Uranus	19.2	84.0 years	17h 14m
Neptune	30.1	164.8 years	16h 06m
Pluto	29.6 to 49.3	247.7 years	6d 9.3h

MOONS OF THE PLANETS

Name	Diameter (km)	Distance From Planet (km from center)	Remarks
EARTH			
Moon	3,476	384,500	Fifth largest moon in solar system; orbital period 27.3d
MARS			
Phobos	21	9,400	Orbital period 7.7h; orbits Mars 3 times while planet turns once
Deimos	12	23,500	Smallest known moon in solar system; orbital period 30.3h
JUPITER			
Metis	40	128,000	At outer edge of Jupiter's ring; orbital period 7h 03m
Adrastea	25	129,000	Just beyond edge of Jupiter's ring; orbital period 7h 08m
Amalthea	170	180,000	Orbital period 11h 57m
Thebe	100	222,000	Orbital period 16h 11m
Io	3,630	422,000	Fourth largest moon in solar system; orbital period 42.5h
Europa	3,140	671,000	Orbital period 3d 13.2h
Ganymede	5,260	1,070,000	Largest moon; mass = 2 x Earth's Moon; orbital period 7.2d
Callisto	4,800	1,885,000	Third largest moon; mass = 1.5 x Moon; orbital period 16.7d
Leda	15	11,110,000	Likely captured asteroid; orbital period 240d
Himalia	185	11,470,000	Likely captured asteroid; orbital period 251d
Lysithea	35	11,710,000	Likely captured asteroid; orbital period 259d
Elara	75	11,740,000	Likely captured asteroid; orbital period 260d
Ananke	30	21,200,000	Captured asteroid; retrograde; orbital period 631d
Carme	40	22,350,000	Captured asteroid; retrograde; orbital period 692d
Pasiphae	50	23,330,000	Captured asteroid; retrograde; orbital period 735d
Sinope	35	23,370,000	Captured asteroid; retrograde; orbital period 758d
SATURN			
Pan	20	134,000	In Encke gap near rings' outer edge; orbital period 13h 51m
Atlas	30	137,000	At outer edge of main ring; orbital period 14h 25m
Prometheus	100	139,000	Orbital period 14h 43m
Pandora	90	142,000	Orbital period 15h 04m
Janus	190	151,000	Shares orbit with Janus; orbital period 16h 41m
Epimetheus	120	151,000	Shares orbit with Epimetheus
Mimas	390	187,000	Orbital period 22h 36m
Enceladus	500	238,000	May have inactive water volcanoes; orbital period 32.9h
Tethys	1,060	295,000	Photograph page 36; orbital period 45.3h
Telesto	25	295,000	Shares Tethys' orbit 60° behind
Calypso	25	295,000	Shares Tethys' orbit 60° ahead

Note: Retrograde moons orbit in the opposite direction from their parent planet's axis rotation.

Name	Diameter (km)	Distance From Planet (km from center)	Remarks
Dione	1,120	378,000	Orbital period 2d 17.7h
Helene	30	378,000	Shares Dione's orbit 60° ahead
Rhea	1,530	526,000	Orbital period 4d 12.4h
Titan	5,150	1,221,000	Mass = 1.8 x Moon; nitrogen atmosphere; orbital period 15.9d
Hyperion	255	1,481,000	Orbital period 21.3d
Iapetus	1,460	3,561,000	Orbital period 79.3d
Phoebe	220	12,960,000	Likely captured asteroid; retrograde; orbital period 550.5d
URANUS			
Cordelia	25	49,800	Voyager 2 discovered 10 inner moons in 1986; orbital period 8.0h
Ophelia	30	53,800	Orbital period 9.0h
Bianca	45	59,200	Orbital period 10.3h
Cressida	65	61,800	Orbital period 11.1h
Desdemona	60	62,600	Orbital period 11.4h
Juliet	85	64,400	Orbital period 11.8h
Portia	110	66,100	Orbital period 12.3h
Rosalind	60	70,000	Orbital period 13.4h
Belinda	70	75,300	Orbital period 14.9h
Puck	155	86,000	Orbital period 18.3h
Miranda	485	129,900	Discovered in 1948; orbital period 33.9h
Ariel	1,160	190,900	Photograph page 39; orbital period 2d 12h
Umbriel	1,190	266,000	Orbital period 4d 4h
Titania	1,610	436,300	Orbital period 8d 17h
Oberon	1,550	583,400	Orbital period 13d 11h
NEPTUNE			
Naiad	60	48,000	Voyager 2 discovered 6 inner moons in 1989; orbital period 7.2h
Thalassa	80	50,000	Orbital period 7.4h
Despina	150	52,500	Orbital period 7.9h
Galatea	160	62,000	Orbital period 10.3h
Larissa	190	73,600	Orbital period 13.2h
Proteus	420	117,600	Orbital period 26.9h
Triton	2,700	354,000	Orbital period 5d 9h; retrograde
Nereid	340	5,510,000	Orbital period 365.2d; very elliptical orbit
PLUTO			
Charon	1,270	19,100	Very large moon for a small planet; orbital period 6d 9.3h

SOURCES

BOOKS

New discoveries keep pushing books about the planets out of date at a rapid pace. It's always a good idea to check the copyright date at the front of a book to ensure that you are reading something fairly recent.

The Grand Tour by Ron Miller and William K. Hartmann (Workman; rev. ed., 1993). Superbly illustrated, factual and up to date, this is the best popular-level book on the solar system I have seen. Highly recommended.

The Universe and Beyond by Terence Dickinson (Camden House; rev. ed., 1992). A generously illustrated introduction to the universe. Half the book is devoted to planets, moons and planets of other stars.

Extraterrestrials by Terence Dickinson and Adolf Schaller (Camden House; 1994). Much of this book is about scientifically plausible worlds of other stars and the life that might inhabit them.

Pale Blue Dot by Carl Sagan (Random House; 1994). A beautifully written and well-illustrated book about the exploration of the solar system. For more advanced readers.

The New Solar System edited by J. Kelly Beatty and Andrew Chaikin (Cambridge; 3rd ed., 1990). A thorough description of planets, asteroids and comets for more advanced readers. Each chapter is written by a world authority. Well illustrated.

Astronomy Today by Eric Chaisson and Steve McMillan (Prentice Hall; 1993). This college astronomy textbook has a very thorough section on the solar system and is a fine reference for the serious student. Other modern college texts offer similar sections, but be sure that you are using a recent edition.

MAGAZINES

Most libraries subscribe to at least one of the following magazines, which offer excellent, up-to-date information.

The Planetary Report, published by The Planetary Society, 65 North Catalina Avenue, Pasadena, CA 91106-2301.

Sky & Telescope, Box 9111, Belmont, MA 02178-9111.

Astronomy, Kalmbach Publishing Co., Box 1612, Waukesha, WI 53187.

POSTERS/VIDEOS/CD-ROMS

For free catalogs of astronomy slides, videos, posters, computer software and more, contact: The Planetary Society, 65 North Catalina Avenue, Pasadena, CA 91106-2301; Astronomical Society of the Pacific, 390 Ashton Avenue, San Francisco, CA 94112; Hansen Planetarium Publications, 1845 South 300 West #A, Salt Lake City, UT 84115; Sky Publishing Corp., Box 9111, Belmont, MA 02178-9111; Kalmbach Publishing Co., Box 1612, Waukesha, WI 53187.

THE AUTHOR

Fascinated by astronomy since childhood, Terence Dickinson remembers reading his first book about the planets at age 8. Today, he is one of North America's leading astronomy writers, with 12 books and more than 1,000 magazine and newspaper articles to his credit.

Mr. Dickinson has received numerous national and international awards for his work, among them the New York Academy of Sciences Book of the Year award and the Royal Canadian Institute's Sandford Fleming Medal. In 1995, in recognition of his achievements in communicating science to the public, he was appointed a Member of the Order of Canada. Asteroid 5272 Dickinson is named after him.

CREDITS

Front and Back Cover Illustrations: David Egge.

p. 4 Terence Dickinson photo; p. 5 NASA photo; p. 6-7 John Bianchi illustration; p. 8-9 John Bianchi illustration; p. 10 (left) David Egge illustration, (right) Terence Dickinson photo; p. 11 David Egge illustration; p.12 NASA photo; p.13 Terence Dickinson photo; p. 14 (left) David Egge illustration, (right) USGS/Phobos 2 photo; p. 15 Space Telescope Science Institute photo; p. 16 NASA photo; p. 17 USGS/NASA photo; p. 18 (left) NASA photo, (right) David Egge illustration; p. 19 NASA photo; p. 20 NASA photo; p. 21 NASA photo; p. 22 NASA photo; p.23 University of Arizona photo; p. 24 David Egge illustration; p. 25 Space Telescope Science Institute photo; p. 26 David Egge illustration; p. 27 David Egge illustration; p. 28 David Egge illustration; p. 29 NASA photo; p. 30 NASA photo; p. 31 NASA photo; p. 32 David Egge illustration; p. 33 Space Telescope Science Institute photo; p. 34 (both) David Egge illustrations; p. 35 David Egge illustration; p. 36 NASA photo; p. 37 NASA photo; p. 38 David Egge illustration; p. 39 NASA photo; p. 40 David Egge illustration; p. 41 (both) NASA photos; p. 42 David Egge illustration; p. 43 David Egge illustration; p. 44 David Egge illustration; p. 45 Space Telescope Science Institute photo; p. 46 David Egge illustration; p. 47 Martin Duncan and Harold Levison photo; p. 48 (both) David Egge illustrations; p. 49 Michael Watson photo; p. 50 Space Telescope Science Institute photo; p. 51 Space Telescope Science Institute photo; p. 52 David Egge illustration; p. 53 David Egge illustration; p. 54 David Egge illustration; p. 55 David Egge illustration; p. 56 David Egge illustration; p. 57 David Egge illustration.

INDEX

Aldrin, Buzz, 12, 58
Alpha Centauri, 4
Andromeda Galaxy, 8
Apollo Moon missions, 12, 58
Ariel, 39
asteroid belt, 9, 22-23
asteroids, 22-23
Barnard's Star, 54
Bonestell, Chesley, 20
brown dwarfs, 52-53
Burroughs, Edgar Rice, 14
Callisto, 30-31
Caloris Basin, 20
Cassini space probe, 36
Cassini's division, 32
Ceres, 22
Cernan, Gene, 12
Charon, 45
Clarke, Arthur C., 36
Comet Shoemaker-Levy 9, 50-51
comets, 46-51
craters, 11, 12-13
Dactyl, 22
Dione, 36
double stars, 54
Earth, 5
  formation of, 10-11
Earth's Moon, 5, 12-13
Einstein, Albert, 53
Enceladus, 32
Europa, 28-29
Gagarin, Yuri, 58
galaxies, 8
Galileo spacecraft, 4-5, 26
Ganymede, 30-31
Glenn, John, 58
Goddard, Robert, 42
greenhouse effect, 18
Halley's Comet, 48
Iapetus, 36
ice
  on Callisto, 30
  on Europa, 28-29
  on Ganymede, 30
  on Mars, 18
  on Mercury, 21
  on Saturn's moons, 36
Ida, 22
Io, 25, 26-27
Jupiter, 24-25, 28, 50-51, 54-55
Kuiper belt, 46-47, 48
Levy, David, 50
life
  on Europa, 28-29
  on Mars, 14-17
Local Group (of galaxies), 8
Lowell, Percival, 14
Mars, 14-17
Mercury, 20
methane, 34-35
Milky Way Galaxy, 8
Mimas, 36
Moon (Earth's), 5, 12-13
  origin of, 12
moons of the planets (see individual names)
  data table, 60-61
  illustration, 6-7
Neptune, 40-41
Olympus Mons, 16-17
Oort cloud, 44, 46
Orion Nebula, 10
Phobos
  photo of, 14
Pioneer 10, 58
Planet X, 46
planetesimals, 10-11
planets (see individual names)
  data table, 59
  of other stars, 54-55
  of pulsars, 56-57
Pluto 44-46
  atmosphere of, 44
  discovery of, 46
  orbit of, 44-45, 47
pulsars, 56-57
Sagan, Carl, 4, 24
Saturn, 32-33
Shoemaker, Eugene, 50, 51
solar system
  illustrations of, 6-9
  origin of, 10-11
  size of, 4
solar wind, 48
Sputnik 1, 58
Sun, 6, 10
  location in galaxy, 8-9
Tethys, 36
Titan, 34-35
Tombaugh, Clyde, 46
Triton, 42-43
  geysers on, 42-43
  origin of, 43
Uranus, 38
Venus, 18-19
volcanoes
  on Io, 26-27
  on Mars, 16-17
Voyager 1, 26-32
Voyager missions, 26-43
water
  on Europa, 28-29
  on Mars, 16
Wells, H.G., 14